Environmental Science, Engineering and Technology

nova science publishers
www.novapublishers.com

Environmental Science, Engineering and Technology

Conserving, Halting and Reinstating Degraded Natural Resources During the UN Decade on Ecosystem Restoration
Rajani Srivastava, M.Sc., PhD (Editor)
Pankaj Kumar, PhD (Editor)
2023. ISBN: 979-8-89113-067-8 (Hardcover)
2023. ISBN: 979-8-89113-140-8 (eBook)

Modeling for Sustainable Development: A Multidisciplinary Approach
Rajendra Kumar, PhD (Editor)
R. C. Singh, PhD (Editor)
Rohit Khokher, PhD (Editor)
Vishal Jain, PhD (Editor)
2023. ISBN: 979-8-89113-015-9 (Softcover)
2023. ISBN: 979-8-89113-056-2 (eBook)

Pesticide Residues: Chemistry, Toxicology and Environmental Impact
David J. Goode (Editor)
2023. ISBN: 979-8-88697-557-4 (Softcover)
2023. ISBN: 979-8-88697-594-9 (eBook)

Emerging Environmental Applications of Nanozymes
Seema Nara, PhD (Editor)
Smriti Singh (Editor)
2023. ISBN: 979-8-88697-552-9 (Hardcover)
2023. ISBN: 979-8-88697-592-5 (eBook)

Atmospheric Aerosols: Properties, Sources and Detection
Binoy K Saikia, PhD (Editor)
2022. ISBN: 979-8-88697-211-5 (Softcover)
2022. ISBN: 979-8-88697-341-9 (eBook)

More information about this series can be found at
https://novapublishers.com/product-category/series/environmental-science-engineering-and-technology/

Eduardo Pavan Korf
Pedro Domingos Marques Prietto
Caroline Müller
and Suéllen Tonatto Ferrazzo
Editors

Environmental Geotechnics

Waste and Geotechnical Stabilization

nova
science publishers
www.novapublishers.com

https://doi.org/10.52305/ZICV4558

NOTICE TO THE READER

Library of Congress Cataloging-in-Publication Data

ISBN: 979-8-89113-651-9 (Hardcover)
ISBN: 979-8-89113-727-1 (eBook)

Published by Nova Science Publishers, Inc. † New York

"The knowledge in soil stabilization is widespread, practiced, and documented in geotechnical engineering. However, searching for and updating new materials as stabilizers for soils and other materials, such as mining tailings stabilization, is necessary. This book brings innovations in the stabilized materials area, both in terms of new methodologies, materials, dosages, and evaluation of the physical-chemical-mechanical-environmental-microstructure behavior of stabilized composites, as well as in the development of geopolymers for applications in distinct areas of geotechnical engineering. It is an important contribution to improve the academic and the practice in the new geotechnical materials development, by incorporating sustainable materials and predicting less generation of environmental impacts."

(Drª Michéle Dal Toé Casagrande)

Contents

Preface

This book addresses the most important and current issues of waste valorization, geotechnical stabilization of different waste types, alternative binders to Portland cement, and metal encapsulation of residues as important strategies for sustainable engineering practice. The development of alkali-activated binders from different wastes (e.g., industrial and agro-industrial wastes, mining tailings, and red ceramic) and their applications are discussed in several chapters. These new alkaline cements and polymers are used in environmental geotechnics to stabilize soils (e.g., expansive soils) and residues (e.g., mining tailings, reclaimed asphalt pavement, and waste foundry sand). In geotechnical stabilization, studies in the literature have focused on the evaluation of mechanical behavior and microstructure. Recently, the leaching of metals and the assessment of sustainability has become a topic of growing interest, which is also presented in this book. In this context, this book highlights the main advantages, challenges, and trends in the use of alternative binders to solve geotechnical environmental problems. It also covers more general aspects of the various soil and waste stabilization methods and their applications.

Chapter 1

Geotechnical Material Stabilization Techniques and Their Application for Different Waste Types

Mariana Krogel
Êmili da Silva Piceta
Daniel Galiano
Caroline Müller
and Eduardo Pavan Korf*
Graduate Program in Environmental Science and Technology,
Universidade Federal da Fronteira Sul, Erechim, RS, Brazil

Abstract

Stabilization is a geotechnical process that includes different techniques and is an old, rudimentary practice that has been improved over time. It is currently recognized as an approach that offers more effective and economical solutions to reduce greenhouse gas (GHG) emissions, a key concern in geotechnical and civil engineering. In particular, the road sector is one of the main contributors to carbon dioxide (CO_2) emissions in the last century, accounting for about 20% of global emissions. Therefore, the purpose of this chapter is to provide a comprehensive review of the literature on the different stabilization techniques used to improve geotechnical and civil engineering properties through the use of wastes from industrial and agricultural activities, among others. In addition, the chapter provides important insights, such as likely

* Corresponding Author's Email: eduardo.korf@uffs.edu.br.

In: Environmental Geotechnics
Editors: E. Pavan Korf, P. Domingos Marques Prietto, C. Müller et al.
ISBN: 979-8-89113-651-9

applications and an overview of life cycle analysis and circular economy and their potential in the context of sustainability. These play a very important role in evaluating the production chain of products, whether hypothetical, under development or existing, with the aim of making the production process as environmentally friendly as possible and conserving natural resources, as well as in assessing the impact of stabilization techniques and reducing CO_2 emissions. The results obtained in the elaboration of this scope show that soil stabilization can be carried out using different methods. These can be divided into three categories: mechanical, granulometric and chemical stabilization. In mechanical and granulometric stabilization, the gradation of the soil is changed, while in chemical stabilization the physical-synthetic properties are changed. It should be remembered that poorly prepared soils for construction, combined with the weathering that occurs over time, will cause damage to structures in the short or long term. Therefore, stabilization techniques play a fundamental role in construction works, requiring their application as a preventive measure and not only as a corrective measure.

Keywords: geotechnical, greenhouse gas, stabilization techniques, sustainability

Soil Stabilization

Soil is the most common and abundant construction material in the Earth's crust, as it is solely responsible for supporting the active loads of any structure (Vizcarra, 2010). It is a component that occurs in a wide variety of environments, such as roads, dams, buildings, landfills, and other engineered structures. Therefore, the knowledge of soil geotechnical properties is essential for the development of projects in terms of safety and economic efficiency (Sharma et al., 2017).

The definition of soil stabilization is described by Winterkorn & Pamukcu (1991) as a set of methods aimed at increasing the resistance of a material by changing one of its properties to improve its behavior. It is also considered as any alternative treatment that supports and stabilizes the soil with the aim of limiting the water absorption capacity and improving the compressibility of the treated soil (Pandey et al., 2023; Soltani et al., 2017). Stabilization is important as a preventive and safety measure because soil has a great variability in its characteristics, and its stabilization cannot be considered only as a corrective measure (Ingles & Metcalf, 1972).

Thus, stabilization corresponds to an artificial treatment of the soil through the use of stabilizing materials with the aim of improving its properties and geotechnical characteristics, that is: increase of resistance, decrease of compressibility and sensitivity, increase or decrease of permeability. In addition, the stabilization process ensures the durability and resistance of the properties in terms of their durability and the action of loads or erosion (Villibor, 1972). These geotechnical techniques alter and improve the condition of the soil and are used where soil replacement is not possible for technical, environmental, or unprofitable reasons (Welsh et al., 2001).

In addition, although stabilization is associated with corrective treatment, it is also a preventive safety measure against adverse conditions that develop during construction or throughout the life of the structure (Ingles & Metcalf, 1973). There are still cases where compaction alone is not sufficient and is used in conjunction with other stabilization methods (Santos et al., 2012). Figure 1 shows a summary flowchart of the three main soil stabilization methods.

There are cases where soils in some regions do not meet the required specifications for use for construction purposes, with problems related to lack of stability, bearing capacity and excessive settlement (Ghadir & Ranjbar, 2018). Therefore, these soils are stabilized using appropriate techniques to improve their resistance properties (Sharma et al., 2020). Soil stabilization can be characterized by three different methods: mechanical, granulometric, and chemical stabilization (Figure 1). Each of these methods has its own mechanism of action but the same application.

An example is the soils in the northern region of Rio Grande do Sul (RS), which are deep and have latosol horizons with a clayey texture (MSRS, 2023). Therefore, they present a double swelling and shrinking behavior, being prone to large volume changes due to variations in water content (Sharma et al., 2020). Due to this behavior, these soils pose serious civil engineering problems, e.g., in the case of infra and suprastructures affected by differential soil settlement, causing cracks and slits in these construction elements (Figure 2) (Krogel et al., 2023), in the case of failed slopes or slopes prone to failure (Figure 3), and in the case of flexible pavements leading to mapped cracks with disintegration (Figure 4a) or disintegration of the pavement only (Figure 4b), differential settlements (Figure 4c), and elevated elevations (Figure 4d), the main cause being related to expanded clay soil, poor design (excessive unforeseen loads), or natural causes (Tanyıldızı et al., 2023).

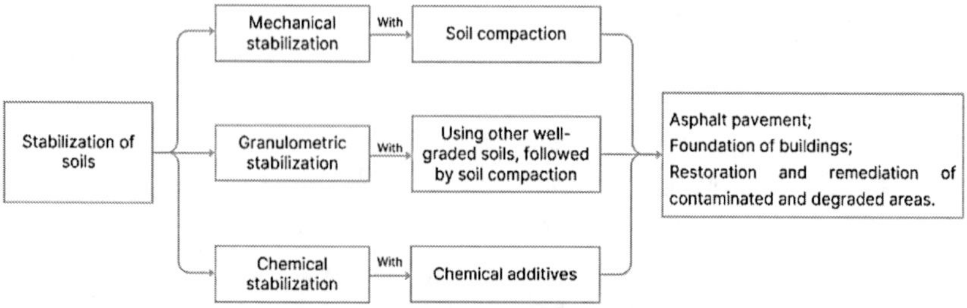

Figure 1. Summary flowchart of stabilization and its applications.

Figure 2. Soil-related problems manifested in structures.

Figure 3. Soil-related problems causing slope failure.

In summary, stabilization of geotechnical materials refers to the process of improving the properties of soils, rocks, or other geotechnical materials to meet specific engineering requirements such as strength, durability, impermeability, and bearing capacity. When applying stabilization techniques to different types of waste, it is important to consider the type of waste, its geotechnical properties, and the stabilization objectives. The types of stabilization, their application to different types of waste, life cycle analysis, and circular economy aspects of waste management are discussed below.

Figure 4. Soil-related problems manifested in flexible pavements.

Mechanical Stabilization

Mechanical stabilization is considered the oldest method in road construction (Medina et al., 2005) and is also referred to as a physical stabilization technique because it is a method that physically changes soil properties and includes compaction, pre-wetting, reinforcement, and electrokinetic treatment (Afrin, 2017; Soltani et al., 2022; Tanyıldızı et al., 2023).

This is done by compacting and rearranging the particles with the aim of minimizing porosity and giving the soil a higher density, because the smaller the volume of voids, the greater the increase in resistance (Mossadeghi-

Björklund et al., 2019). Stability is achieved by the compaction energy exerted on the soil. This inhibits the infiltration of water and the erosion it causes, and increases the density, durability, compactness, and consequently the mechanical resistance of the soil (Santos et al., 2012).

However, the application of this method has its limitations, namely: (i) they are generally not independent techniques that need to be periodically improved with chemical stabilization; (ii) in situ quality control is insufficient or time-consuming; (iii) the result is not reliable for critical soils (Ikeagwuani & Nwonu, 2019).

Particle Size Stabilization

Granulometric stabilization is a technique commonly used on construction sites and between pavement layers, primarily because it is a technique that generally complements other stabilization techniques (Bernucci et al., 2006). An example of such a technique is the use of hydraulic macadam, the most common material used on the first Brazilian highways. It is a granular layer of coarse, natural or crushed aggregates whose voids on the pavement are filled with fine aggregates and, in the specific case of hydraulic macadam, bound by water.

In granulometric stabilization, the objective is to obtain a well-graded material with a limited amount of fine particles by intimately homogenizing two or more soils and then compacting them (Vizcarra, 2010). The increase in mechanical strength used in this method results from the contact between the grains of the larger particles and the filling of the voids by the grains of the finer particles (Villibor, 1982). It aims to modify the properties of the soil by adding or removing particles, creating a homogeneous mixture of two or more soils that are subsequently compacted (Medina & Motta, 2015). In other words, soils with different particle size fractions are mixed.

Each soil type has different particles in size and fraction, and the determination of the properties of the particles and their occurrence allows obtaining the particle size distribution. It is important to consider that the particle size distribution of each soil type is influenced by the type of particles and the optimal compaction by stabilization. Therefore, the particle size stabilization process must produce a final product that is suitable for use and complies with the normative specifications established for each specific application (Medina & Motta, 2015; Vizcarra, 2010).

Chemical Stabilization

Chemical stabilization is performed by adding materials, traditional or not (Barman & Dash, 2022; Rivera et al., 2020). It has the potential to change the structure of the soil through the addition of chemical materials (Vizcarra, 2010). The goal is to increase stability by increasing the particle size of the soil material and decreasing the plasticity index and expansion and contraction potential. This reaction takes place between the stabilized material and the minerals present in the soil (Fondjo et al., 2021; Makusa, 2013).

The stabilizer can be traditional materials, such as bitumen, Portland cement, lime, pozzolan, and industrial products (Huang et al., 2021; Rivera et al., 2020). Or by non-traditional materials, such as industrial wastes (see Chapter 3) and even alkali-activated binders (Karami et al., 2021; Levandoski et al., 2023; Pelissaro et al., 2023) (see Chapters 8, 9, and 10). In this method, it is necessary to perform laboratory tests for in situ application to predict the effectiveness of the stabilized materials and to determine the type of stabilizer and its dosage.

According to Lambe & Michaels (1954), it is important for a chemical stabilizer to fulfill three essential mechanisms to satisfactorily modify the soil, namely: (i) hydrophobicity of the particles by removing adsorbed water as well as soil surface loading; (ii) increase of interparticle repulsion resulting from an increase in the surface electronegativity of the particles or the thickness of the water membrane; and (iii) facilitation of interparticle cohesion. To meet the requirements of each mechanism, specific chemical agents are used, such as impregnating agents, dispersants, and aggregates/flocculants.

Materials Used in Chemical Stabilization

A variety of materials can be used in stabilization: Cement, lime, and fly ash are considered conventional additives, while sulfonated oils, polymers (see Chapter 2), salt, and enzymes are considered nontraditional additives, as are additives from byproducts such as Portland cement powder, lime kiln dust, and slag (Barman & Dash, 2022; Bernucci et al., 2006). They can be used in two different ways, depending on the objective: i) in the case of a significant consolidation of the soil, mass proportions between 5 and 8% are usually used, and the resulting material is called soil cement; ii) in the case of a partial improvement of properties, mainly workability associated with some increase in bearing capacity, low proportions between 2 and 4% are used, in this case called a mixture of improved soil and cement (DNIT, 2022).

The traditional method with added lime is the most effective, which is why it is widely used, especially in expansive soils (Barman & Traço, 2022; Herrin & Mitchell, 1961). The main factors that make it the most widespread and effective application are cation exchange, flocculation and agglomeration, pozzolanic reaction, and carbonation, which are responsible for changing soil structure by reducing plasticity and solidification (Arabi & Wild, 1986).

The reaction mechanisms begin with the assumption of exothermic reactions when quicklime (CaO) is added to a clayey soil and, in the presence of water, helps to reduce its content due to the heat of hydration, as shown in reaction equation 1 (Rogers & Glendinning, 2000). When dissolved in water, the hydrated lime decomposes into Ca^{2+} and OH^- ions, as shown in reaction equation 2 (Barman & Traço, 2022).

$$CaO + H_2O \rightarrow Ca(OH)_2 + \text{heat of hydration} \quad (1)$$

$$Ca(OH)_2 \rightarrow Ca^{2+} + 2OH^- \quad (2)$$

Calcium ions (Ca^{2+}) replace the monovalent cations present in soil minerals and cause an increase in the attractive force between particles. As a result, the thickness of the diffuse double layer of clay particles decreases, which reduces swelling changes in the presence of water (Elkady, 2016). These calcium ions are usually exchanged for metal ions associated with the clay structure through the process of cation exchange.

The hydroxyl ions (OH^-) released by hydrated lime create a high pH ($\geq$12.5) in which silica and clay can dissolve (Silva et al., 2022). Once silica and alumina are in the soil structure, they react with water and lime to form calcium silicate hydrate (C-S-H) and calcium aluminate hydrate gel (C-A-H), which subsequently crystallize and bind the structure. These reactions are referred to as pozzolanic reactions (Rogers & Glendinning, 2000).

Although the stabilization mechanisms of lime and cement are almost identical (Barman & Traço, 2022; Estabragh et al., 2012), cement is used because of its easy availability and rapid strength gain (Zhang et al., 2022). However, it should preferably be processed in a plant, but it can be mixed on the road for small quantities or for roads with low traffic (Bernucci et al., 2006). The stabilization mechanism consists of hydration, cation exchange, flocculation, and pozzolanic reaction (Barman & Traço, 2022). The difference is that hydration starts immediately when the soil is mixed with cement and water. And the development of strength can be done by the cement hydration (Figure 5a) and by the products of the pozzolanic reaction (Figure 5b).

During cement hydration (Figure 5a), bonds form between the hydrating cement and soil particles, forming a highly resistant mass. The formation of cement bonds reduces the potential for leaching of calcium hydroxide when the soil is exposed to seasonal wetting and drying cycles or when groundwater moves through the stabilized soil (Bhattacharja et al., 2003; Prusinski & Bhattacharja, 1999).

Furthermore, in the pozzolanic reaction (Figure 5b), the reactions proceed slowly, further strengthening the amended soil, reducing its plasticity, and improving its gradation. It is postulated that calcium ions combine with silica and alumina dissolved in the soil structure to form C-S-H and C-A-H (Bhattacharja et al., 2003; Cyr et al., 2013; Silva et al., 2022). The stabilization technique used must be able to maintain the high pH of the soil for a long period of time so that pozzolanic reactions become a significant factor in improving the resistance and durability of the system. Otherwise, the system may be susceptible to leaching, resulting in a drop in pH, interruption of pozzolanic action, and return of the soil to its original state (Barman & Dash, 2022; Prusinski & Bhattacharja, 1999).

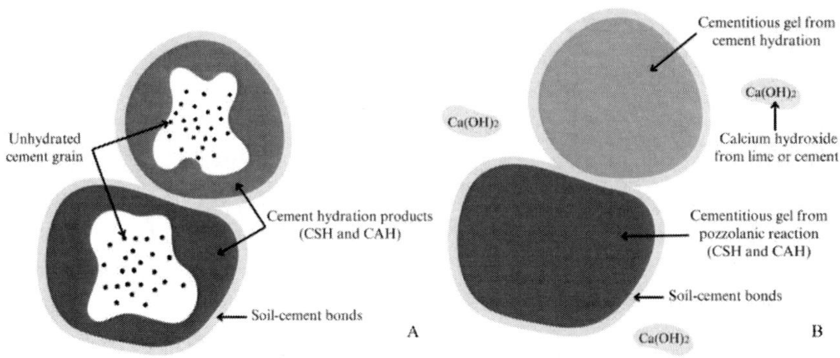

Figure 5. Mechanisms of strength development by cement. Adapted from Prusinski & Bhattacharja (1999).

Although the traditional method is the widely used and standardized method, the construction industry depends on the global reserves of limestone and natural aggregates, and the related criticism refers to the scarcity of supply in the future (Ioannidou et al., 2017). In addition, the production of Portland cement and the extraction of limestone used for both soil stabilization and building construction account for about 40% of the sector's global CO_2 and other GHG emissions due to the operation of clinker kilns alone (IEA, 2018, 2020; Taylor et al., 2006), taking into account that cement production alone contributes about 7 to 8% of global CO_2 and other GHG emissions (IEA, 2020; Okoye et al., 2015).

This shows that civil engineering and the development of industries seriously impede progress in reducing carbon emissions (Komkova & Habert, 2023). Therefore, the current interest in the use of alternative materials, the use of non-traditional stabilizers in soil stabilization, can be successfully carried out, improving the properties of expansive subgrade materials, reducing the environmental impact and the overall cost of construction. (Fondjo et al., 2021; Gupta & Kumar, 2017; Kumar et al., 2007). It also contributes to the reduction of carbon emissions, natural resource depletion, and relocation of economic activities (Kassa et al., 2020; Komkova & Habert, 2023).

Geopolymers from byproducts that include rice husk ash, ground granulated blast furnace slag, plastic waste, synthetic fibers, sugarcane bagasse ash, cow dung ash, fly ash, bituminous, silica fume, geotextiles, and tissues, among others, can be used in soil stabilization (Pelisser et al., 2023; Rivera et al., 2021). The application of alkali-activated binders may be appropriate for clayey soils or for dynamic loading systems such as highways and railroads (Barman & Traço, 2022) (see Chapters 8 and 9). This is because the binders have a more linear chemical structure and thus possess adhesive properties and promote better soil behavior (MacKenzie et al., 2006).

The reaction mechanisms responsible for the physical, chemical, and mineralogical changes in the treated soils are aimed at modifying the plasticity of the soil or solidifying it (Davidson et al., 1965). Modification occurs through flocculation and cation exchange activity, while consolidation occurs through a pozzolanic reaction (Salehi & Sivakugan, 2009). However, care must be taken in the presence of organic matter in the soil as it interferes with the process of pozzolanic reactions, as organic matter tends to form a film around the additive grains that prevents their hydration, thus reducing the resistance of the treated soil or even its efficiency (Barman & Dash, 2022).

Application of Stabilization Methods in Waste

When applying stabilization methods to different types of waste, it is important to consider the type of waste and its geotechnical properties (Al-Taie et al., 2019). Chemical stabilization involves applying a specific mixture to the soil to improve its properties. Traditional methods are still the most commonly used, including the use of lime and cement. The difference is that the use of cement improves the resistance of the soil through pozzolanic reactions, effectively improving the permanent deformation of silty and sandy soils, while lime supports plasticity to reduce the expansion of clayey soils (Abu-Farsakh et al., 2015).

The application can aim to stabilize soils or even use waste in construction materials to create a solid and cohesive matrix from by-products (Fatahi & Khabbaz, 2015). An example of this is the study by Korf et al., (2020), which investigated the use of Portland cement in clayey soils with different compaction conditions as an impermeable barrier to prevent seepage of acidic leachate in this soil.

In addition, chemical stabilization technology has the ability to encapsulate toxic substances and reduce contaminant migration (Eighmy et al., 1997), which is commonly referred to as stabilization/solidification. An example of this application is the research conducted by Câmara (2023), in which paint booth residues were encapsulated during the manufacture of concrete blocks to prevent the leaching of pollutants contained in the residues into the environment. And in the study of Jovelino et al., (2020), who used the stabilization technique in the treatment of solid wastes from chemical laboratories to allow easy and appropriate disposal, promising results were obtained in the retention of heavy metals (cadmium and nickel above 98%), suggesting that stabilization is a viable option to reduce the amount of heavy metals in solid wastes from chemical laboratories.

Mechanical and granulometric stabilization can be accomplished by different processes, but are usually performed together to significantly improve soil behavior (Villibor, 1972). While mechanical stabilization aims to improve the physical properties of the soil, such as its resistance, bearing capacity, and durability, through the addition of materials such as cement, lime, chemical stabilizers, or other materials of a scientific nature, granulometric stabilization aims to improve the granulometric distribution of the soil, i.e., to adjust the proportion of different particle sizes present in the soil (Arrivabeni, 2017).

An example of this is the study by Lima et al., (2020), who used these techniques together on a sandy laterite soil using ceramic waste from construction waste. The results showed that the use of these wastes in combination with stabilization techniques proved to be viable for use in pavements, as they significantly increased the bearing capacity of the soil and made it suitable for use in pavements by stabilizing lateritic soils sandy.

And also, in the study of Cavalcanti et al., (2022), who proposed the improvement of compacted soils by adding alternative materials such as stone powder and polypropylene fibers through mechanical and granulometric stabilization. The results showed the improvement of a soil with low compressive strength and proved that the applied physical and granulometric stabilization methods are a technically feasible alternative and can be used in base courses of landfills and road pavements, among others.

In the work of Morais et al., (2023), only granulometric stabilization was used and the reuse of discarded foundry sand in mixtures with crushed material commonly used in road construction was evaluated, where it was applied as a granular layer on the road pavement. Despite this difference, all of the tests met the guidelines defined by DNIT and showed the potential to be used both as a sub-base and as a base for pavements, provided that appropriate compaction energy is applied.

Life Cycle Analysis and Circular Economy in the Context of Sustainability in Civil Construction

The concept of sustainability is discussed from different angles. According to Boff (2016), sustainability can be defined as the effort to restore the balance of the earth and ecosystems so that the "common house" remains habitable. In contrast to the production and consumption patterns that cause environmental degradation, the only solution is a global alliance in the scientific, technical, and especially social spheres (Carta da terra, 2000; Zhang et al., 2022).

As the world population grows, waste production and construction activity follow this expansion (Al-Subari et al., 2023). Therefore, sustainable development has become a fundamental approach in this sector, especially in earthworks, as geotechnical engineering is a significant contributor to resource and energy consumption (Abreu et al., 2008; Rocha et al., 2016). Therefore, quantifying the potential environmental impacts in these sectors is necessary to develop alternative solutions to promote sustainable growth.

Life cycle analysis (LCA) and the circular economy (CE) model are important allies in the context of sustainability. Considering that they are complementary techniques when used together at different stages of the production chain, the combination of LCA and CE allows product manufacturers to develop and compare strategies with full environmental assessments that ensure a balanced ecological system. According to Mannan & Al-Ghamdi (2022):

> While implementing LCA in CE evaluation, LCA can be utilized to compliment CE in different stages. First, for any hypothetical product or process (macro, micro, or regional level) the pros and cons of CE can be analyzed through LCA. Second, possibly better alternatives for the entire life cycle can be identified through LCA once the limitations of a circular model are recognized during the first stage. Finally, LCA continuously sets goals and aims for improving circularity in real-life scenarios. Therefore, combining CE and LCA ensures mobilizing business with proven benefits for the environment and society. (Mannan & Al-Ghamdi, 2022, p. 153)

The ONU (2015) "proposes seventeen Sustainable Development Goals (SDGs) based on the "2030 Agenda," and each of the goals presents its specific targets, with actions focused on people, planet, and prosperity.

Among the seventeen SDGs, SDG 11 - Sustainable Cities and Communities - stands out, more specifically Target 11.6: By 2030, reduce the negative per capita environmental impacts of cities, paying particular attention to air quality and the management of municipal and other wastes (ONU, 2015). The use of life cycle assessment and circular economy methods allows for the creation of a system-conscious economy that contributes to the goals proposed by SDG 11 and enables changes in consumption and production patterns, making the use of all available natural resources sustainable (ONU, 2015; Webster, 2021).

Life Cycle Analysis

Infrastructure works and construction projects cause environmental impacts for which geotechnical engineering is mostly responsible. Therefore, studies that evaluate the sustainability of production processes in this field are fundamental. According to Ferrazzo (2023), there are two broad categories within geotechnical engineering that encompass research focused on sustainability: those related to promoting the use of alternative materials and those related to the development of structures and tools that help evaluate sustainability.

In the context of soil improvement, one can observe the increasing application of life cycle analysis in geotechnical engineering research aimed at assessing sustainability (Al-Subari et al., 2023; Rocha et al., 2016). Ferrazzo (2023) lists examples of LCA application in his study, such as containment structures (Damians et al., 2015), soil improvement methods (Raymond et al., 2021), remediation of contaminated areas (Visentin et al., 2022), and traditional and alternative paving (Da Rocha et al., 2022).

In soil stabilization, all known products, as well as new products used in research and in the field, are being developed to meet existing needs (e.g., the current demand for energy efficiency in the era of global warming). From the moment a product is developed, it has a "life" that spans all stages of its production cycle, such as raw material extraction, manufacturing, packaging, transportation, consumption, and disposal. All methods used in each phase of product production generate some type of environmental impact, whether it is the unconscious consumption of natural resources, the emission of substances, the generation of contaminating waste, and the improper disposal of packaging, the latter usually practiced by the end user (Rebitzer, 2004).

LCA is a waste management method that can be added to institutional goals to use resources more efficiently and reduce costs. According to Hunkeler et al., (2003), the introduction of life cycle analysis in organizations leads to a change in all decisions made in the production phases and directly affects the overall impact that the product will have during its life cycle, since life cycle analysis must be integrated at all levels and, to be effective, must remain flexible.

ISO 14040 (2009a) defines LCA as a "cradle-to-grave" systemic approach, meaning that the entire life cycle of products is assessed, from the sourcing of raw materials used in manufacturing, through the production process, use and post-use treatment, to final disposal, including recycling and reuse, if applicable, focusing on environmental aspects and potential environmental impacts that arise during the material production process, in the application and post-use process. Application, avoiding emerging concerns (Raymond et al., 2017). Figure 6 shows the structure of life cycle analysis, which consists of four phases.

As shown in Figure 6, the four phases that make up the structure of a life cycle analysis are interconnected. In each of these phases, the objectives are defined and the data to be analyzed are collected so that critical points in the product life cycle can be identified and initiatives to improve production processes can be taken on this basis.

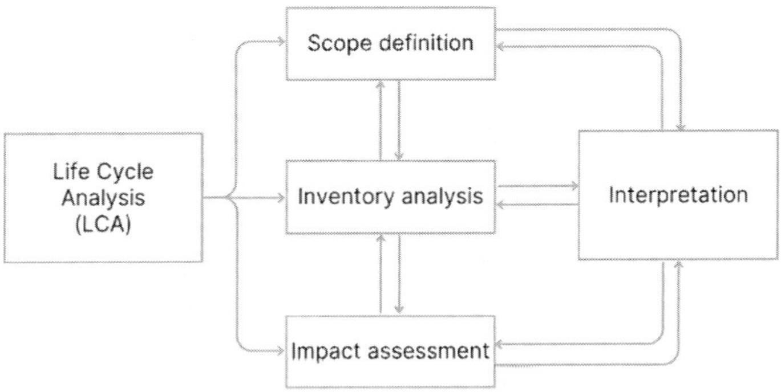

Figure 6. Structure of life cycle analysis.

The first step in a life cycle assessment is to define the scope. In this first moment, the study is delineated in terms of system boundaries and level of detail. In this phase, some criteria defined by ISO 14044 (2009b) are considered, such as the product system targeted by the study, the quality parameters, the way the data are interpreted, the environmental impacts to be assessed, the reporting format to be produced, and the intended use of the results. It is important to emphasize that LCA can vary greatly in depth and scope depending on the study and its objectives.

The second phase is the inventory analysis. Based on the objectives established in the first phase, qualitative and/or quantitative data collection begins. The collected data can be measured, calculated or estimated and are used to quantify the inputs and outputs of an elementary process (ISO 14044, 2009b). In this phase, data validation is performed in addition to energy flow calculations and correlation of the data with the elementary processes.

At the end of the second phase, the third phase begins, the so-called impact assessment. According to ISO 14040 and ISO 14044 (2009a, b), a carefully planned significance analysis is carried out on the basis of the potential environmental impacts already identified in the scope definition phase and the results obtained in the inventory analysis phase. of the potential environmental impacts, which also provides information on the life cycle of the products. This phase also determines whether or not the objectives established for the study have been met, and how the scope can be modified if this is the case.

Finally, the interpretation phase, the fourth stage of the LCA structure, begins. In this stage, the considerations highlighted in the Inventory Analysis and Impact Analysis stages are analyzed together, with the goal of providing results that are consistent with the objective and scope, and presenting conclusions that explain the limitations and make recommendations. The ISO 14040 (2009a) standard indicates that the results of the last phase of LCA are based on a relative approach and therefore it is important to reflect on the results. In addition, those responsible for decision-making within institutions may consider the results of this final phase as conclusions and recommendations for actions to promote changes in the life cycle of products that follow the objectives outlined during the development of the LCA.

The growing demand for solutions in the field of waste, especially municipal solid waste, and the need for new alternatives for its management is noteworthy. In this context, life cycle analysis shows a huge potential for application in the management of solid waste and industrial and agro-industrial by-products, since its unique methodology makes it possible to identify and follow different pathways for the efficient and energetic management of waste (Weitz et al., 1999; Mannan & Al-Ghamdi, 2022). Thus, promoting progress with a positive impact on the environment

Finally, it should be highlighted that LCA can be considered as an innovative eco-tool, since its scientifically based methodology predicts the saving of natural resources and the reduction of environmental impacts. According to Motta & Issberner (2018), the application of life cycle analysis enables the transition from a linear economy to a circular economy.

Circular Economy

Albuquerque & Callado (2010) explains how humans have exploited natural resources in the past without caring about the damage caused by this uncontrolled exploitation. The assumed linear production model that contributed to industrial development, especially after World War II, was based on the extraction of natural resources, which were considered abundant and unlimited in nature, and their processing, production, use, and disposal. Natural resources were viewed as inexpensive and easily accessible commodities, and this view encouraged the uncontrolled use of these resources, whether in construction or other sectors, as well as the generation of large amounts of waste (Barauna et al., 2018; Ellen MacArthur Foundation, 2015). Considering that the planet's resources are limited and finite, as well as its capacity to absorb all the waste and pollution generated, it is inevitable that we reflect on the need to transform the linear economic and production model, with the aim of reduce the use of natural resources and alleviating their scarcity. In this context, the circular economy is in contrast with the idea of the linear economy.

Stahel (2016) elaborates on how the adherence to circular economy systems in the global economy becomes an inevitable fact in terms of sustainability that guarantees human life and its quality. As a concept that runs completely counter to the end-of-life concept of the linear economy, the circular economy proposes the reduction, reuse, recovery, and recycling of materials and energy, maximizing the value of each phase of a product's life. This remanufacturing of material goods has positive impacts on the environment and the population, as new jobs are created in addition to energy savings, reduced consumption of natural resources, and lower waste generation (Geisendorf & Pietrulla, 2017; Stahel, 2016; Start & Go, 2018).

Under the concept of circular economy, wastes generated during the production and consumption of products become resources that can be used, especially in the geotechnical field. An example of waste management from the perspective of the circular model is the reference study developed in Portugal by the Grande Porto Intermunicipal Waste Management Service (LIPOR). The approach adopted by the institution consists in a strategy of recovery, treatment and containment of municipal waste based on multi-material valorization, organic valorization and energy valorization. The waste collected by LIPOR goes through a specific separation process that starts with the citizen as a waste producer, whose actions have a direct impacts on the rest of the chain. Once deposited and collected, the waste goes through several selection stages that lead to the verification of the resulting products and resources. An example of a resource resulting from this process is the use of organic material as fertilizer, replacing the use of chemical fertilizers (Ferreira, 2015; Start & Go, 2018).

Another example is the challenges faced by civil infrastructure when encountering low-grade geomaterials (e.g., mining wastes, sludges, dredged sediments) that, to remedy, are usually removed or landfilled, leading to a new challenge in waste management (Mavroulidou et al., 2021). Therefore, chemical stabilization with alkali-activated binders helps in this process of remediation of low-grade soils by avoiding the accumulation of solid wastes both in the soil when it is removed and when wastes and by-products are used in the development of the binder and promotes the circular economy of solid waste (Duxson & Provis, 2008). Such as the research of Levandoski et al.

(2023), who evaluated the stabilization of iron ore tailings with an alkali-activated binder produced from sugarcane bagasse ash, hydrated eggshell lime, and sodium hydroxide solution.

The circular economy has already been introduced or adopted in some countries such as China and the European Union. Despite the countless benefits, there are many challenges in implementing the circular model. Berardi & Dias (2018) point to challenges such as product costs, which in the linear economy model divide the costs of external care of the product among market participants or are imposed on the public sector, while in a linear economy model these costs would be internalized; consumer behavior, which must change its relationship with products with the aim of their functionality rather than ownership, is identified as an even greater challenge than cost, as it involves cultural and social factors that are difficult and time-consuming to change. In terms of implementation in production chains, the circular model requires partnerships and collaborations in maintenance and manufacturing, as the complexity of operations and logistics increases (Berardi & Dias, 2018).

Final Considerations

Given this overview, it is clear that the stabilization process can be applied not only to soils but also to wastes, contributing to sustainable waste management and reducing environmental impacts. In this context, circular economy and life cycle assessment are two fundamental approaches, both of which aim to reduce environmental impacts, promote the efficient use of resources, and minimize waste generation. Incorporating waste stabilization techniques can further strengthen a strategy to promote sustainability in natural resource and waste management.

Chemical stabilization focuses on the use of traditional materials such as cement and lime, with an emphasis on specific categories of materials. In contrast, for materials that are considered waste, the focus is on specific types such as fly ash, slag, or construction and demolition waste. However, there is a research gap related to the comprehensive evaluation of a wider variety of waste types and the study of how the different chemical and physical compositions of these wastes affect the properties of the stabilization method.

One research idea is to develop uniform criteria for selecting stabilization methods for geotechnical materials based on the properties of the waste and site conditions. These criteria can help engineers select the most appropriate technique for a given situation and may include analyzes of durability and resistance to different environmental conditions and over time. Another aspect is to establish clear guidelines and regulations for the application of geotechnical waste stabilization techniques. This would help ensure the safety, effectiveness, and applicable regulations for stabilization procedures in different contexts.

Acknowledgments

The author thanks the National Council for Scientific and Technological Development (CNPq) and Federal University of Fronteira Sul (UFFS) for financial support. M. Krogel and C. Müller are grateful to the Coordination for the Improvement of Higher Education Personnel (CAPES) and D. Galiano and E. P. Korf to the CNPq for their fellowships.

References

Abreu, D. G., Jefferson, Braithwaite, P. A., Chapman, D. N. & Student, P. D. (2008). Why is sustainability important in Geotechnical Engineering? *Geocongress: geosustainability and geohazard mitigation*. https://doi.org/10.1061/40971(310)102.

Abu-Farsakh, M., Dhakal, S. & Chen, Q. (2015). Laboratory characterization of cementitiously treated/stabilized very weak subgrade soil under cyclic loading. *Soils and Foundations, 55(3)*, 504–516. https://doi.org/10.1016/j.sandf.2015.04.003.

Afrin, H. (2017). A Review on different types soil stabilization techniques. *International Journal of Transportation Engineering and Technology, 3(2)*, 19. https://doi.org/10.11648/j.ijtet.20170302.12.

Albuquerque, J. L. & Callado, A.L.C. (2010). *Gestão Ambiental E Responsabilidade Social: Conceitos, Ferramentas e Aplicações.* ISBN 8522457727, 9788522457724. São Paulo: Atlas, 1st ed. 336 p.

Al-Subari, B. L., Hilal, A. & Ekinci, A. (2023). Life cycle assessment of soil stabilization using cement and waste additives. *Construction and Building Materials, 403*, 133045. https://doi.org/10.1016/j.conbuildmat.2023.133045.

Al-Taie, A., Hussein, A. & Ali, A. (2019). Uma revisão sobre a estabilização de solos expansivos usando diferentes métodos. *Revista de Engenharia Geotécnica, 6*, 32-40. https://doi.org/10.3759/JOGE.V6I3.3354.

Arabi, M. & Wild, S. (1986). Microstructural composites development in cured soil-lime. In *Journal of Materials Science*. Vol 21. doi: https://doi.org/10.1007/BF01145514.

Arrivabeni, B. S. (2017). *Alternativas para estabilização granulométrica e química de solo de estradas florestais*. Dissertação apresentada à Universidade Federal de Viçosa.

ABNT – Associação brasileira de normas técnicas. (2009a). *Gestão Ambiental – Avaliação do ciclo de vida – Princípios e estrutura.* (ABNT NBR ISO 14040: 2009a). ABNT. https://www.normas.com.br/.

ABNT – Associação brasileira de normas técnicas. (2009b). *Gestão Ambiental – Avaliação do ciclo de vida – Requisitos e Orientações.* ABNT NBR ISO 14044: 2009b). ABNT. https://www.normas.com.br/.

Barauna, D., Souza, S., Trein, F. A. & Razera, D. L. (2017). Design para a sustentabilidade na economia de materiais: uso de resíduos no desenvolvimento de produtos. In *Mix Sustentável, 3(3)*, 113-122. https://doi.org/10.29183/2447-3073.MIX2017.v3.n3.113-122.

Barman, D. & Dash, S. K. (2022). Stabilization of expansive soils using chemical additives: A review. In *Journal of Rock Mechanics and Geotechnical Engineering* (Vol. 14, Issue 4, pp. 1319–1342). Chinese Academy of Sciences. https://doi.org/10.1016/j.jrmge.2022.02.011.

Berardi, P. & Dias, J. M. (2018). O mercado da Economia Circular: Como os negócios estão sendo afetados pelo modelo que substitui o linear e como serão ainda mais a médio e longo prazos. In *GV Executivo, 17(5)*, 34-37. https://doi.org/10.12660/gvexec.v17n5.2018.77340.

Bernucci, L. L. B., Motta, L. M. G. da, Ceratti, J. A. P. & Soares, J. B. (2008). *Pavimentação asfáltica: formação básica para engenheiros.* https://repositorio.usp.br/directbitstream/0d316eee-ef3b-4fb0-bdf7-fbd2bc7b0863/Bernucci-2008-Pavimentacao%20asfaltica.pdf.

Bhattacharja, S., Bhatty, J. I. & Todres, H. A. (2003). Stabilization of clay soils by Portland cement or lime - A critical review of literature. 60 p. Accessed May 2023. Available from: https://www.cement.org/docs/default-source/cement-concrete-applications/sn2066.pdf?sfvrsn=5f54fdbf_2.

Boff, L. (2016). *Sustentabilidade: O que é – O que não é.* Editora Vozes Limitada. ISBN: 8532656102, 9788532656100.

Câmara, C. M. (2023). *Utilização de resíduo de cabine de pintura para produção de blocos de concreto intertravado.* Universidade Federal da Fronteira Sul.

Cavalcanti, E., Silva, A., Lins, C., Amorim, S. F. & Lins, E. A. M. (2022). *Avaliação de diferentes métodos de estabilização mecânica físico-granulométrica dos solos da Formação Cabo para aplicação em obras de engenharia.* In: proceedings in XX COBRAMSEG. Accessed June 2023. Available from: https://proceedings.science/cobramseg-2022/trabalhos/avaliacao-de-diferentes-metodos-de-estabilizacao-mecanica-fisico-granulometrica?lang=pt-br.

Cyr, M., Trinh, M., Husson, B. & Casaux-Ginestet, G. (2014). Effect of cement type on metakaolin efficiency. *Cement and Concrete Research, 64*, 63–72. https://doi.org/10.1016/j.cemconres.2014.06.007.

Da Rocha, C. G., Saldanha, R. B., Tonini de Araújo, M. & Consoli, N. C. (2022). Social and environmental assessments of eco-friendly pavement alternatives. *Construction and Building Materials, 325*, 126736. https://doi.org/10.1016/j.conbuildmat.2022.12673.

Damians, I. P., Bathurst, R. J., Adroguer, E. G., Josa, A. & LloretL, A. (2015). Environmental assessment of earth retaining wall structures. *Environmental Geotechnics, 4(6)*, 415–431. https://doi.org/10.1680/jenge.15.00040.

Davidson D. T., Demirel T. & Handy R. L. *Soil Pulverization and Lime Migration in Soil-Lime Stabilization*. In Highway Research Record 92, Highway Research Board, National Research Council, Washington D.C., 1965, pp. 103–126.

DNIT – Departamento Nacional de Infra-estrutura de Transportes. (2022). DNIT 143/2022 - ES. Pavimentação-base de solo-cimento-Especificação de serviço. Rio de Janeiro, 2022.

Det Udomsap, A. & Hallinger, P. (2020). A bibliometric review of research on sustainable construction, 1994–2018. In *Journal of Cleaner Production* (Vol. 254). Elsevier Ltd. https://doi.org/10.1016/j.jclepro.2020.120073.

Diemer, F., Rambo, D., Specht, L. P. & Pozzobon, C. E. (2008). Propriedades geotécnicas do solo residual de basalto da região de Ijuí/RS. In *Teoria e Prática na Engenharia Civil*, Rio Grande - Rs, v. 12, p. 25-36.

Duxson P. & Provis JL (2008). Designing precursors for geopolymer cements. *Journal of the American Ceramic Society, 91(12)*, 3864–3869. https://doi.org/10.1111/j.1551-2916.2008.02787.x.

Eighmy, T., Crannell, B., Butler, L., Cartledge, F., Emery, E., Oblas, D., Krzanowski, J., Eusden, J., Shaw, E. & Francis, C. (1997). Estabilização de metais pesados na combustão de resíduos sólidos urbanos de depurador seco utilizando fosfato solúvel. *Ciência e Tecnologia Ambiental, 31(11)*, 3330-3338. https://doi.org/10.1021/es970407c.

Elkady, T. Y. (2016). The effect of curing conditions on the unconfined compression strength of lime-treated expansive soils. *Road Materials and Pavement Design, 17(1)*, 52–69. https://doi.org/10.1080/14680629.2015.1062409.

Ellen Macarthur Foundation. (2015). Towards a Circular Economy: Busines Rationale for an Accelerated Transiton. In *Ellen MacArthur Foundation*. Accessed May 2023. Available from: https://archive.ellenmacarthurfoundation.org/assets/downloads/publications/TCE_Ellen-MacArthur-Foundation_26-Nov-2015.pdf.

Estabragh, A. R., Namdar, P. & Javadi, A. A. (2012). Behavior of cement-stabilized clay reinforced with nylon fiber. *Geosynthetics International, 19(1)*, 85–92. https://doi.org/10.1680/gein.2012.19.1.85.

Fatahi, B. & Khabbaz, H. (2015). Influence of chemical stabilisation on permeability of municipal solid wastes. *Geotechnical and Geological Engineering, 33(3)*, 455–466. https://doi.org/10.1007/s10706-014-9831-y.

Ferrazzo, S. T. (2023) *Resíduo de areia de fundição estabilizado com ligante álcali- ativado de cinza de bagaço de cana-de-açúcar e cal de casca de ovo: Desempenho mecânico, microestrutura, lixiviação e sustentabilidade*. Universidade Federal do Rio Grande do Sul.

Ferreira, A. V. M. (2015). *Aplicação de uma ferramenta de Análise de Ciclo de Vida no planeamento da gestão integrada de resíduos urbanos: estudo de caso da Lipor*. Universidade Aberta.

Fondjo, A. A., Theron, E. & Ray, R. P. (2021). Stabilization of expansive soils using mechanical and chemical methods: A Comprehensive review. *Civil Engineering and Architecture, 9(5)*, 1289–1294. https://doi.org/10.13189/cea.2021.090503.

Geisendorf, S. & Pietrulla, F. (2017). The circular economy and circular economic concepts-a literature analysis and redefinition. *Thunderbird International Business Review, 60(5)*, 771-782. https://doi.org/10.1002/tie.21924.

Ghadir, P. & Ranjbar, N. (2018). Clayey soil stabilization using geopolymer and Portland cement. *Construction and Building Materials, 188*, 361–371. https://doi.org/10.1016/j.conbuildmat.2018.07.207.

Gupta, D. & Kumar, A. (2017). Performance evaluation of cement-stabilized pond ash-rice husk ash-clay mixture as a highway construction material. *Journal of Rock Mechanics and Geotechnical Engineering, 9(1)*, 159-169. https://doi.org/10.1016/j.jrmge.2016.05.010.

Herrin, M. & Mitchell, H. (1961). Lime-soil mixtures. *Highway Research board bulletin.*

Huang, J., Kogbara, R. B., Hariharan, N., Masad, E. A. & Little, D. N. (2021). A state-of-the-art review of polymers used in soil stabilization. In *Construction and Building Materials*. Vol. 305. doi: https://doi.org/10.1016/j.conbuildmat.2021.124685.

Hunkeler, D., Saur, K., Stranddorf, H., Rebitzer, G., Finkbeiner, M., Schmidt, W-P., Jensen, A. A., Cristiansen, K. (2003). *Life Cycle Management*. Pensacola, Florida: Setac Press, 120 p. Accessed May 2023. Available from: https://www.researchgate.net/publication/299470360_Life_cycle_management.

IEA. (2020). Special Report on Clean Energy Innovation. In *Energy Technology Perspectives 2020*. OECD Publishing, Paris. https://doi.org/10.1787/ab43a9a5-en.

IEA. (2018). Global Status Report: Towards a zero-emission, efficient and resilient buildings, and construction sector. International Energy Agency. Accessed April 2023. Available from: https://www.iea.org/reports/2018-global-status-report.

Ikeagwuani, C. C. & Nwonu, D. C. (2019). Emerging trends in expansive soil stabilisation: A review. In *Journal of Rock Mechanics and Geotechnical Engineering*. Vol. 11, Issue 2, pp. 423–440. Chinese Academy of Sciences. doi: https://doi.org/10.1016/j.jrmge.2018.08.013.

Ingles, O. G. & Metcalf, J. B. (1972). *Soil Stabilization: Principles and Practice*. London, England: Butterworth and Company Publishers Ltd. ISBN 0-470-42742-6. 374 p.

Jovelino, J. R., Brito, A. L. F. de, Muniz, A. C. S. & Da Silva, P. P. (2020). Estabilização/solidificação: caracterização, classificação e aplicação no tratamento e gerenciamento de resíduos sólidos de laboratórios químicos. *Holos Environment, 20(4)*, 539–558. https://doi.org/10.14295/holos.v20i4.12410.

Karami, H., Pooni, J., Robert, D., Costa, S., Li, J. & Setunge, S. (2021). Use of secondary additives in fly ash-based soil stabilization for soft subgrades. *Transportation Geotechnics, 29*, 100585. https://doi.org/10.1016/j.trgeo.2021.100585.

Kassa, R. B., Workie, T., Abdela, A., Fekade, M., Saleh, M. & Dejene, Y. (2020). Soil stabilization using waste plastic materials. *Open Journal of Civil Engineering, 10(1)*, 55–68. https://doi.org/10.4236/ojce.2020.101006.

Komkova, A. & Habert, G. (2023). Optimal supply chain networks for waste materials used in alkali-activated concrete fostering circular economy. *Resources, Conservation and Recycling, 193*, 106949. https://doi.org/10.1016/j.resconrec.2023.106949.

Korf, E. P., Prietto, P. D. M., Silveira, A. A., Ulsen, C., Consoli, N. C. & Bragagnolo, L. (2020). Modifications in the chemical and mineralogical composition of compacted mature residual soil submitted to the percolation of acidic leachates. In *Soils and Rocks, 43(1)*, 159–167. https://doi.org/10.28927/SR.431159.

Krogel, M., Langoski, M. & Ferrarese, R. G. (2023). Analysis of cracks in family houses buildings in the city of Erechim – RS. In *XIX International Conference on Building Pathology and Constructions Repair.* ISBN 978-65-86819-09-0. doi: https://doi.org/10.4322/CINPAR.2023.001.

Kumar, A., Walia, B. S. & Bajaj, A. (2007). Influence of fly ash, lime, and polyester fibers on compaction and strength properties of expansive soil. *Journal of Materials in Civil Engineering, 19(3)*, 242-248. https://doi.org/10.1061/(ASCE)0899-1561 (2007)19:3(242).

Lambe, T. W. & Michaels, A. (1954). Altering soil properties with chemicals. *Chemical Engineering, 32(6)*, 488-492.

Levandoski, W. M. K., Ferrazzo, S. T., Bruschi, G. J., Consoli, N. C. & Korf, E. P. (2023). Mechanical and microstructural properties of iron mining tailings stabilized with alkali-activated binder produced from agro-industrial wastes. *Scientific Reports, 13)*, 15754. https://doi.org/10.1038/s41598-023-42999-x.

Lima, J. S. P., Vicente, G. O., Fagundes, L. P., Cabral, C. L. B., da Cruz, F. A. & Tashima, M. M. (2020). Physical stabilization of a sandy lateritic soil using ceramic wastes. *Revista Materia, 25(4)*, 1–11. https://doi.org/10.1590/S1517-707620200004.1176.

Ioannidou, D., Meylan, G., Sonnemann, G. & Habert, G. (2017). Is gravel becoming scarce? Evaluating the local criticality of construction aggregates. *Resources, Conservation and Recycling, 126*, 25–33. https://doi.org/10.1016/j.resconrec.2017.07.016.

MacKenzie, K. J. D., Brew, D., Fletcher, R., Nicholson, C., Vagana, R. & Schmücker, M. (2011). Advances in understanding the synthesis mechanisms of new geopolymeric materials. Novel Processing of Ceramics and Composites: Ceramic Transactions Series, Volume 195. https://doi.org/10.1002/9781118144114.ch19.

Makusa, G. P. (2013.). State of the art review of soil stabilization methods and materials in engineering practice. Accessed May 2023. Available from: https://www.diva-portal.org/smash/get/diva2%3a997144/FULLTEXT01.pdf.

Mannan, M. & Al-Ghamdi, S. G. (2022). Chapter 9 - Complementing circular economy with life cycle assessment: Deeper understanding of economic, social, and environmental sustainability. In: *Circular Economy and Sustainability*, Elsevier. Pages 145-160. ISBN 9780128198179. doi: https://doi.org/10.1016/B978-0-12-819817-9.00032-6.

Mavroulidou, M., Gray, C., Gunn, M. J. & Pantoja-Muñoz, L. (2022). A study of innovative alkali-activated binders for soil stabilisation in the context of engineering sustainability and circular economy. *Circular Economy and Sustainability, 2(4)*, 1627–1651. https://doi.org/10.1007/s43615-021-00112-2.

Medina, J. de. & Motta, L. M. G. da. (2015). *Mecânica dos Pavimentos*. 3th ed. Rio de Janeiro: Editora Interciência. 640 p.

MossadeghiBjörklund, M., Jarvis, N., Larsbo, M., Forkman, J. & Keller, T. (2019). Effects of compaction on soil hydraulic properties, penetration resistance and water flow patterns at the soil profile scale. In *Soil Use and Management.* 35, 367 - 377. https://doi.org/10.1111/sum.12481.

Motta, W. H. & Issberner, L-R. (2018). Rumo à economia circular: Qual o papel da ACV. In: *Congresso brasileiro sobre gestão do ciclo de vida*, 6., 2018, Brasília. Anais do VI Congresso Brasileiro sobre Gestão do Ciclo de Vida. Brasília: Ibict. p. 92-97.

MSRS – Museu de solos do Rio Grande do Sul. (2023). Available from: https://www.ufsm.br/museus/msrs/unidade-de-solos.

Okoye, F. N., Durgaprasad, J. & Singh, N. B. (2015). Mechanical properties of alkali activated fly ash/Kaolin based geopolymer concrete. *Construction and Building Materials*, 98, 685–691. https://doi.org/10.1016/j.conbuildmat.2015.08.009.

Organização das Nações Unidas no Brasil – ONU BR. *A Agenda* 2030. 2015.

Pandey, B.K., Shukla, C., Sillanpää, M. & Shukla, S. K. (2023). A systematic review on application of electrokinetics in stabilization and remediation of problematic soils. *Innovative Infrastructure Solutions, 8, 226*. https://doi.org/10.1007/s41062-023-01186-7.

Pelissaro, D. T., Bruschi, G. J., Pavan Korf, E. & Dalla Rosa, F. (2023). Rice husk ash as an alternative soluble silica source for alkali-activated metakaolin systems applied to recycled asphalt pavement stabilization. *Transportation Geotechnics, 39*, 100940. https://doi.org/10.1016/j.trgeo.2023.100940.

Pelisser, G., Ferrazzo, S. T., Mota, J. D., dos Santos, C. P., Pelisser, C., Rosa, F. D. & Korf, E. P. (2023). Rice husk ash-carbide lime as an alternative binder for waste foundry sand stabilization. *Environmental Science and Pollution Research, 30(14)*, 42176–42191. https://doi.org/10.1007/s11356-022-25117-8.

Prusinski, J. R. & Bhattacharja, S. (1999). Effectiveness of Portland cement and lime in stabilizing clay soils. *Transportation research record, 1652(1)*, 215-227. https://doi.org/10.3141/1652-28.

Raymond, A. J., S. M., Pinkse, M. A., Kendall, A. & Dejong, J. T. (2017). Life-cycle assessment of ground improvement alternatives for the treasure Island, California, Redevelopment. In *Geotechnical Frontiers 2017*. ISBN 9780784480434. https://doi.org/10.1061/9780784480434.037.

Raymond, A. J., Dejong, J. T., Kendall, A., BlackburnL, J. T. & Deschampsr, R. (2021). Life cycle sustainability assessment of geotechnical ground improvement methods. In *Journal of Geotechnical and Geoenvironmental Engineering, 147(12)*, 04021161. https://doi.org/10.1061/(ASCE)GT.1943-5606.0002646.

Rebitzer, G., Ekvall, T., Frischknecht, R., Hunkeler, D., Norris, G., Rydberg, T., Schmidt, W. P., Suh, S., Weidema, B.P. & Pennington, D.W. (2004). Life cycle assessment Part 1: Framework, goal and scope definition, inventory analysis, and applications. *Environment International*, v. 30, n. 5, p. 701-720, 2004. https://doi.org/10.1016/j.envint.2003.11.005.

Rivera, J. F., Orobio, A., Mejía De Gutiérrez, R. & Cristelo, N. (2020). Clayey soil stabilization using alkali-activated cementitious materials. *Materiales de Construccion, 70(337)*, e211. https://doi.org/10.3989/mc.2020.07519.

Rivera, J., Coelho, J., Silva, R., Miranda, T., Castro, F. & Cristelo, N. (2021). Compressed earth blocks stabilized with glass waste and fly ash activated with a recycled alkaline cleaning solution. *Journal of Cleaner Production, 284*, 124783. https://doi.org/10.1016/j.jclepro.2020.124783.

Rocha, C. G., Passuello, A., Consoli, N. C., Quiñónez Samaniego, R. A. & Kanazawa, N. M. (2016). Life cycle assessment for soil stabilization dosages: A study for the Paraguayan Chaco. *Journal of Cleaner Production, 139*, 309–318. https://doi.org/10.1016/j.jclepro.2016.07.219.

Rogers, C. D. F. & Glendinning, S. (2000). Lime Requirement for Stabilization. *Transportation Research Record*, 1721(1), 9-18. https://doi.org/10.3141/1721-02.

Santos, M. N., Ghavami, K. & Pinto, A. R. A. G. (2012). *Análise do efeito da estabilização mecânica em matrizes de terra.* http://www.Chacaracolinadosventos.org/res/site54452/res440132_.

Sharma, A. & Sharma, R. K. (2020). Sub-grade characteristics of soil stabilized with agricultural waste, constructional waste, and lime. *Bulletin of Engineering Geology and the Environment, 80*, 2473–2484. https://doi.org/10.1007/s10064-020-02047-8/Published.

Sharma, L. K., Umrao, R.K., Singh, R., Ahmad, M. & Singh, T. N. (2017). Geotechnical characterization of road cut hill slope forming unconsolidated geo-materials: A case study. *Geotechnical and Geological Engineering* 35, 503–515. https://doi.org/10.1007/s10706-016-0093-8.

Silva, P. B., Ângulo, S. C., Motta, R. & Bernucci, L. L. B. (2023). Pozzolanic reaction of recycled cementitious and red ceramic-waste aggregates with hydrated lime for pavement base layer. *Journal of Materials in Civil Engineering, 35(2)*, 04022422. https://doi.org/10.1061/(asce)mt.1943-5533.0004579.

Soltani, A., Deng, A., Taheri, A. & O'Kelly, B. C. (2022). Intermittent swelling and shrinkage of a highly expansive soil treated with polyacrylamide. *Journal of Rock Mechanics and Geotechnical Engineering, 14(1)*, 252–261. https://doi.org/10.1016/j.jrmge.2021.04.009.

Soltani, A., Taheri, A., Khatibi, M. & Estabragh, A. R. (2017). Swelling potential of a stabilized expansive soil: a comparative experimental study. *Geotechnical and Geological Engineering* 35, 1717–1744. https://doi.org/10.1007/s10706-017-0204-1.

Stahel, W. R. (2016). The circular economy. In *Nature*, New York, v. 531, n. 7595, p. 435-438.doi: https://doi.org/10.1038/531435a.

Taylor, M. & Tam, C. (2006). *Energy Efficiency and CO_2 Emissions from the Global Cement Industry.* Accessed May 2023. Available from: https://www.researchgate.net/publication/237232402_Energy_Efficiency_and_CO2_Emissions_from_the_Global_Cement_Industry.

Tanyıldızı, M., Uz, V. E. & Gökalp, İ. (2023). Utilization of waste materials in the stabilization of expansive pavement subgrade: An extensive review. *Construction and Building Materials*. 132435. https://doi.org/10.1016/j.conbuildmat.2023.132435.

Villibor, D. F. Estabilização Granulométrica ou Mecânica. (1972). São Carlos: Gráfica EESC – USP. Accessed May 2023. Available from: http://repositorio.eesc.usp.br/server/api/core/bitstreams/62f2e3c7-1914-41e1-ba3f-9283c525358a/content.

Visentin, C., Da Silva Trentin, A. W., Braun, A. B. & Thomé, A., (2021). Social life cycle assessment of the nanoscale zero-valent iron synthesis process for application in contaminated site remediation. *Environmental Science and Pollution Research, 29*, 21603–21620. https://doi.org/10.1007/s11356-021-17319-3.

Vizcarra, G. O. C. (2010). *Aplicabilidade de Cinzas de Resíduo Sólido Urbano para Base de Pavimentos*. Pontifícia Universidade Católica do Rio de Janeiro.

Webster, K. (2021). A Circular Economy Is About the Economy. Circular Economy and Sustainability. 1, 115–126. https://doi.org/10.1007/s43615-021-00034-z.

Weitz, K., Barlaz, M. A., Ranjithan, R. S., Brill, D., Thorneloe, S. A. & Ham, R. K. (1999). Life Cycle Management of Municipal Solid Waste. In *The International Journal of Life Cycle Assessment,* 4, 195-201. doi: https://doi.org/10.1007/BF02979496.

Welsh, E. V., Warren, J. & Lukas, R. (2001). Ground Improvement Technical Summaries: Volume 2. Earth Engineering & Sciences, Inc. United States. Department of Transportation. *Federal Highway Administration*. FHWA-SA-98-086 R.

Winterkorn, H. & Pamukcu, S. (1991). *Soil Stabilization and Grouting*. In Foundation Engineering Handbook. pp 317–378. https://doi.org/10.1007/978-1-4615-3928-5_9.

Zhang, B., Zhu, H., Feng, P. & Zhang, P. (2022). A review on shrinkage-reducing methods and mechanisms of alkali-activated/geopolymer systems: Effects of chemical additives. *Journal of Building Engineering, 49*, 104056. https://doi.org/10.1016/j.jobe.2022.104056.

Chapter 2

Performance of Mining Tailings Stabilized with Polymers and the Applicability in Geotechnical Engineering

Michéle Dal Toé Casagrande[1,*]
Giovanna Monique Alelvan[2]
and Rodrigo Cesar Pierozan[3]
[1]Department of Civil and Environmental Engineering, University of Brasília, Brasília, DF, Brazil
[2]Department of Transportation and Geotechnical Engineering, Federal University of Minas Gerais, Belo Horizonte, MG, Brazil
[3]Department of Civil Engineering, Federal Institute of Education, Science and Technology of Rondonia, Porto Velho, RO, Brazil

Abstract

Managing the substantial volume of waste generated by mining activities represents one of the foremost challenges the mining industry faces. This challenge stems from the necessity of constructing secure structures employing materials of intricate geotechnical properties, such as mining tailings. The term 'tailings' denotes the residual material resulting from the ore beneficiation process, which holds limited economic value for the mining sector. In this context, reinforcement and stabilization techniques play a pivotal role in enhancing the geotechnical characteristics of the tailings. This chapter comprehensively evaluates the physical, chemical, mechanical, environmental, and microstructural behavior of various mining tailings (including iron and gold ore tailings) in their pure state and after stabilization with polymeric solutions. Alternative polymer-

* Corresponding Author´s Email: mdtcasagrande@unb.br.

In: Environmental Geotechnics
Editors: E. Pavan Korf, P. Domingos Marques Prietto, C. Müller et al.
ISBN: 979-8-89113-651-9

based soil stabilization solutions are also under investigation, given their potential for implementation in mining activities. The study further assesses the enhancement of these composites by incorporating polypropylene fibers. The investigation delves into several critical parameters, including polymer content, granulometry, curing time, specific mass, and water content. Material characterization encompassed grain size distribution, specific unit mass, and compaction curves. The mechanical analysis comprised unconfined compressive strength, direct shear tests, water retention, cyclic triaxial repeated loads (resilience module), laboratory traffic simulation, and stiffness modulus at small strains. Mineralogical characterization was done through X-ray diffraction (DRX) and X-ray fluorescence spectrometry (FRX). Microstructural analysis employs X-ray Microtomography, Scanning Electron Microscopy, and Optical Microscopy, while chemical analysis assesses pH values and thermal performance via Thermogravimetric Analysis. The test results indicate the viability of using this polymer-based stabilizer to repurpose mining tailings in geotechnical structures and complementary alternative solutions. Polymer-based tailings stabilization proves effective in static and cyclical stress scenarios, and adding polymeric fibers substantially enhances the material's ductility. This influence extends to the attenuation of stiffness and the stabilization of residual strength in the polymer-tailings composite. In summary, the findings support using polymers in treating mining tailings as a promising alternative for enhancing their properties, enabling their application in engineering projects such as tailings piles, embankments, and surface coatings for unpaved roads.

Keywords: new geotechnical materials, iron tailings, gold ore tailings, polymers, stabilization, alternative solutions

Introduction

The mining industry is pivotal in the global economy, supplying essential raw materials for various sectors. However, one of the most pressing challenges within this industry is managing the substantial volume of waste generated during mining activities. This challenge arises due to the need to construct secure structures using materials of high geotechnical complexity. Additionally, the construction of waste disposal sites must be carefully synchronized with the operational schedule of mining plants, with a primary objective of maximizing the efficiency of ore production—the industry's primary economic output.

A significant component of this mining waste is "tailings," which comprises the residue left behind after the ore beneficiation process. Tailings are often characterized by their high moisture content and slurry-like consistency upon exiting the beneficiation plant. Traditionally, tailings have been disposed of in tailings dams, a method that has been widely employed in the industry. However, recent years have seen the Brazilian mining sector grappling with substantial environmental disasters due to the catastrophic failure of these tailings dams (see Chapter 10). This has prompted the industry to actively seek alternative waste disposal methods that are structurally less complex and offer greater control over environmental impact.

A novel solution, known as the "dry stacking" of compacted filtered tailings, has emerged in response to these challenges (Alshawmar & Fall, 2023; Cao et al., 2022). This innovative approach aims to significantly reduce moisture content through filtration processes, resulting in dewatered tailings that can be compacted and stored in piles efficiently (Pan et al., 2022). In this context, it becomes crucial to employ reinforcement and stabilization techniques to enhance the geotechnical characteristics of these tailings, making them suitable for engineering applications (Gao et al., 2023). Furthermore, addressing logistical challenges during rainy periods, which can disrupt mineral production and extraction processes, is paramount to maintaining consistent operational efficiency within mining complexes.

In this academic study, a comprehensive literature review is undertaken to investigate the potential of stabilizing mining tailings with polymeric solutions and enhancing their properties through the incorporation of polypropylene fibers. The research seeks to evaluate the influence of various factors, including polymer content, particle size distribution, curing time, specific gravity, and water content, on the performance of these tailings. Furthermore, the study encompasses detailed characterizations of these materials, based on previous works, encompassing grain size distribution analysis, specific unit mass measurement, compaction curve analysis, mineralogical analysis, microstructural analysis, and chemical analysis. The overarching objective is to determine the viability of utilizing polymer-stabilized mining tailings in geotechnical engineering applications, thereby presenting a sustainable and environmentally conscious solution to the challenges associated with mining waste management.

Literature Review

To deal with the treatment of mining tailings from a geotechnical perspective, it is essential to list the contributions that Soil Mechanics, Rock Mechanics, and Geology can make. According to the Committee of Rock Mechanics of the Geological Society of America, rock mechanics is a theoretical and applied science that studies the mechanical behavior of rocks. Although there was some prior knowledge about rocks, it was only recognized as a discipline in engineering programs in 1960.

One of the basic precepts of this science is the clear distinction between rocks and soils, given mainly by the properties of internal cohesion (forces that bind the particles together) and tensile strength. In this context, rocks can be defined as solid materials made up of one or more minerals that occur on the earth's crust in the form of large masses or fragments. Although geotechnical engineers are interested in the mechanical response, evaluating the genesis and the minerals that compose it is essential for interpreting the behavior.

For this reason, the classification of rocks is related to their origin: igneous, when formed from the crystallization of magma; metamorphic, formed from a mineralogical adaptation to pressure and temperature conditions; and sedimentary, formed from the accumulation of sediments, microorganisms, and chemical precipitation.

When rocks are exposed to the atmosphere after they have been formed, they are subject to forces and reagents that are different from those in their original environment. The action of these factors leads to a fragmentation of the rock material, reducing its resistance and transforming it into soil. This process is called weathering, which can be physical, chemical, or biological.

Weathering is influenced by factors such as rock type, mineralogical structure, topography, climate, and the process length. Drawing a parallel with mining tailings, it can be said that the ore processing process is a type of artificial weathering since it subjects the rock material to physical processes - such as mechanical excavations, vibration efforts, wetting and drying cycles, and thermal variations - and chemical processes due to the disposal associated with water and acidic environments (Smuda et al., 2014; Zhang et al., 2023).

On the other hand, metamorphic rocks with a high quartz content are less susceptible to weathering. It can, therefore, be said that gold ore tailings, which are rich in quartz, only undergo physical weathering during mineral processing. This finding is interesting because it highlights that this is a fragmented material with no apparent cohesion since there have been no chemical effects to create bonds (cementation) between the particles (Pierozan et al., 2023).

From the perspective of soil mechanics, much can be said about tailings. (Lambe & Whitman, 1969) pointed out that the nature of soils is distinguished from other materials by their particulate structure and that, for this reason, the science that evaluates the stress-strain behavior of these materials could be called Particulate Mechanics.

These particles interact with each other by direct contact of the grains or by electromagnetic forces. In the case of granular soils, contact interactions are the majority; in clay soils, electromagnetic interactions prevail, depending on the individual structure of each particle and the environment in which they are inserted.

Various authors widely discuss clays' crystalline structure (Nakayenga et al., 2021). In general, clay minerals have two types of characteristic structures: octahedral, in which six oxygens or hydroxyls surround the aluminum, iron, or magnesium atom, or tetrahedral, in which the silicon atom occupies the central position surrounded by four oxygens or hydroxyls.

The arrangement of these minerals generates clay particles with negative charges on the faces and positive charges on the edges. This ionization is generated by adsorption, dissociation, and isomorphic substitution, which cause electromagnetic forces of attraction or repulsion between the particles.

More broadly, according to Lambe & Whitman (1969), the interaction between particles depends on four forces: externally applied, contact, attraction, and repulsion. The first is related to the contact surface and the deformability of the particles and is reflected in effective stresses.

The contact forces are mineral-mineral, mineral-air, and mineral-water interactions. In the case of granular soils, mineral-mineral contact is easily observed, which is not the case in cohesive soils. This contact will only exist in well-flocculated, moisture-free, and cemented soils since the ions and water adsorbed by the minerals generate repulsion forces in this type of contact.

The forces of attraction are due to Van der Waals forces, also known as secondary valence forces. Their characteristic is that they are weaker and are susceptible to modification in the event of a change in the external force.

Finally, the forces of repulsion are electrostatic in nature are preceded by confinement and are very sensitive to changes in the material's structure. They follow the variation in the electrical constant, the hydrated ions' size, the Electrical Double Layer (ECL) thickness, the pH, and the absorption of the anions. However, temperature, ionic valence, and electrolyte concentration vary in the opposite direction.

In this context, the arrangement of the particles will be guided by the forces of attraction or repulsion if external forces are disregarded. If the forces of attraction are predominant, the particles will be arranged in a non-oriented manner, forming a structure known as “flocculated.” On the other hand, if repulsive forces prevail, the particles will be directed in a structure known as “dispersed.”

This crystalline configuration becomes even more complex in soils and tailings, multiphase media where the fluid part directly affects the transmission of forces between the particles, and thus the mechanical behavior. This is an interesting finding from the point of view of stabilizing tailings with polymers because, in the case of water-soluble polymers, there are simultaneous reactions between particles-particles, particles-water, particles-polymer (Galjak et al., 2020; Krishna et al., 2021; Wayal et al., 2021). This dynamic contact between them will depend on the external forces applied, often assessed in the literature by varying the structure of the samples, such as by varying the dry specific mass.

Considering all the above, other points from the literature review will be addressed: interfacial phenomena in tailings, soil stabilization, and polymers applied to geotechnical engineering.

Interfacial Phenomenon in Tailings

Understanding the interfacial relationships in tailings is necessary to establish a parallel with the previous considerations and gain a deeper understanding of the relationships between the tailings and the polymers. This area of study in Mining Engineering is necessary to understand the flotation stage of the ore beneficiation process. The theoretical basis for the phenomena that occur in the flotation stage will first be briefly explained. These concepts will then be expanded to the object of study.

As with soils, tailings are multiphase materials of solid, liquid, and gaseous phases. Phases are also considered to be regions of space where the chemical composition and physical and mechanical properties are equivalent (Cao et al., 2022; Okewale & Grobler, 2023).

The interface between the phases is the boundary layers with a colloidal dimension (1 nm), which have different chemical and physical properties than the other components. There are five types of interfaces: solid-solid, solid-liquid, solid-gas, liquid-liquid, and liquid-gas.

As mentioned above, tailings are fragments generated by fragmenting the parent rock. At this stage, new surfaces are formed by breaking high-energy bonds, so they react spontaneously with the environment to lower their energy level.

The fine particles generated during this stage adhere to each other through electrostatic and Van der Waals forces. For this reason, it is important to assess the reactivity of the surface of these particles with the fluids used since the entire process of obtaining the ore takes place by the wet method (Castillo et al., 2023; Manz, 2024).

At these interfaces, adsorption phenomena occur due to the tendency to interact with other chemical species in the surrounding fluid. Adsorption can be chemical in nature, causing the formation of new compounds on the mineral surface, or physical in nature, where electrostatic forces are predominant. The interactions between minerals and liquids result in an imbalance of charges, which can generate an excess of positive or negative charges, which can be caused by three mechanisms: ionic crystals, surface hydration, or isomorphic substitution.

Briefly, the mechanism of ionic crystals occurs when ionic species on the surface of minerals interact with a polar liquid. This interaction leads to the migration of ions from the solid to the solution and from the solution to the solids (adsorption). This movement occurs until the ionic concentration product reaches electrochemical equilibrium. This mechanism then shows that in contact with polar liquids, minerals can behave like ionic crystals with different degrees of solubility.

The hydration mechanism occurs when the surface of a hydrophilic mineral interacts with water molecules. As a result, the acidic and basic groups are dissociated depending on the pH of the suspension. This creates a negatively charged surface (ionization of acidic groups in an alkaline environment) or a positively charged surface (ionization of basic groups in an acidic environment). The intensity of the surface charge is related to the degree of acidity or alkalinity of the surface group and the solution (Cota et al., 2022; Nekhoul et al., 2022).

Finally, isomorphic substitution is related to some mineral species' natural charge imbalance, especially silicates. This allows these minerals to make isomorphic substitutions in their crystal structure. In general, the most common is the replacement of Si^{4+} by Al^{3+}, with electrical neutrality being re-established by monovalent cations. However, in contact with water, these ions dissociate, leaving the surface of the mineral with an excess of negative charge.

The electrical properties existing at the interface of these materials are studied based on the theory of the Double Electrical Layer (DCE). According to this theory, ions near a surface give rise to DCE, which is a layer with an excess of positive charge, and another with an excess of negative charge, which as a whole is neutral.

All of this theoretical basis is a reference for understanding the flotation stage of beneficiation, which is the stage that gives rise to the tailings investigated here. The flotation stage is part of one of the concentration methods, especially for finer grain sizes, which aims to increase the grade of the exploited ore.

This stage is carried out in equipment that inserts air into the mineral pulp. This mixture then has three phases that interact with each other: the solid mineral particles, the liquid water present in the pulp, and the air, which is broken up into bubbles. Due to the difference in density, this last phase tends to rise to the surface. Along the way, the bubbles collide with the mineral particles, and those that have a greater affinity with the apolar phase (generally hydrophobic) are then adsorbed onto the surface of the bubbles. A chute then collects the flotation material.

However, the challenge of this stage is that 90% of minerals are hydrophilic, i.e., they have an affinity for water. It is, therefore, necessary to induce a differentiating property in some of these particles so that they no longer have an affinity with water and separation is possible. This induction of hydrophobicity is done using chemical reagents.

Within the chemical reagents class, a wide variety of collectors stand out. These compounds are surfactants, i.e., they adsorb onto mineral particles, forming a film to make them hydrophobic. These collectors can be anionic, cationic, or non-ionic. Generally, anionic collectors are used in gold ore flotation, especially potassium amyl xanthate (PAX) (Oliveira Junior, 2011). Some characteristics of this collector are worth mentioning to draw a parallel with the polymers studied here:

- It is highly soluble in water;
- Adsorption is practically only chemical in nature. This is why a conditioning time is needed for the surface compound to form;
- Should be used in slightly alkaline environments;
- The speed of degradation increases with temperature;
- The speed of xanthate decomposition increases with the acidity of the medium.

Understanding the interfacial phenomena that occur during this stage from the collectors is used to evaluate polymer composites. Several authors have already taken this approach to evaluate composites on a microstructural scale, such as (Jiang et al., 2023), who evaluated interfacial phenomena in reinforced iron ore tailings to understand the mechanical behavior under different reinforcement conditions.

Stabilization Solutions for Geomaterials

In their natural state, some geomaterials do not have the minimum properties required for use in geotechnical structures. For this reason, soil improvement techniques can increase resistance, reduce compressibility, achieve permeability requirements, and, in some cases, reduce liquefaction potential.

The stabilization technique covers solutions with chemical additives that react with the natural material or mix materials with complementary particle size distributions to form a more stable structure (see Chapter 1). The reinforcement technique aims to form composites with inclusions of materials with a tensile strength to complement the behavior of the granular materials, enabling a wide range of engineering applications, including slope stabilization, retaining wall construction, and pavement design. These technologies can be used individually or in combination, maximizing the benefits inherent to the techniques.

The main advantage of using these materials lies in substituting conventional construction materials, such as sand and gravel, which may be associated with significant environmental impacts due to the need for natural deposit extraction. Furthermore, many of the new technologies have the characteristic of better-quality control of the products, allowing for a better prediction of the behavior of the final structure, even in the design phase.

When it comes to the use of tailings in geotechnical structures, it is known that improvement or reinforcement solutions are required so that they can act as materials for such applications. These solutions can come from mixing the tailings with another stabilizing material or through methods such as mechanical stabilization.

Due to the shortage of natural material products and environmental restrictions, studying alternative products for stabilization has become compelling. As this type of material does not have traditional technical specifications or a wide range of uses, studies are required to evaluate the efficiency of its application.

Stabilization solutions are physical, mechanical, or chemical processes that seek to adjust the properties of materials to the required performance levels. Physical stabilization is related to grain size correction, which can be achieved by adding fibers.

Mechanical stabilization can be carried out using compaction or granulometric correction. In the first case, the structure of the geotechnical material is altered by applying an external force, which increases the contact between the grains. This reduction in the volume of voids means that displacements during the useful life of the structures are minimized. This solution is widely used for collapsible soils with a metastable structure in their natural state (Khodabandeh et al., 2023; Zimbardo et al., 2020).

On the other hand, particle size correction involves mixing two or more types of material to adjust the particle size curve. This means there will be a better distribution of fractions, allowing for greater interlocking and contact between the grains and, thus, better resistance.

However, particle size stabilization may not be enough for geotechnical materials to meet the requirements for use in engineering works. On the other hand, due to the unavailability of complementary materials in the region, the method becomes unfeasible.

Chemical stabilization, therefore, becomes an attractive option from a technical and economic point of view. This type of improvement is related to treatment using chemical processes. In the field of soils, the traditional stabilizing agents are cement, lime, fly ash, and bituminous materials.

The chemical stabilization of clay soils presents a unique understanding due to the interactions that take place between the matrix and the stabilizer. The concepts studied in interfacial phenomena are essential for understanding the mechanisms. Four chemical reactions take place during the chemical stabilization of clay soils: (i) cation exchange, (ii) agglomeration/flocculation, (iii) hydration, and (iv) pozzolanic reactions. The cation exchange process generates physical and mechanical changes in the soil structure and can generally take place over long periods. The mechanism consists of the interaction of ions released by the stabilizer (calcium ions) with the metal ions of the material itself (see Chapter 3).

Polymers Applied in Geotechnical Engineering

The term polymer is of Greek origin and means the joining together of many parts (polys). Each part is called a manomer and the number of places where the molecules connect is called the functionality. When these various units react to form a polymer chain, this is called polymerization. It is important to note that polymers are different from plastics. A 'polymer' represents the pristine material derived from the polymerization process, seldom employed in its pure state, while 'plastic' results from additives.

Polymers can be classified according to their occurrence, with natural ones being those that exist in nature (e.g., protein, cellulose, starch, rubber) and artificial ones being those generated in the laboratory (e.g., polyethylene, Styrofoam). Concerning the method used to obtain them, they can be classified as addition, when obtained by adding a single monomer; copolymers, when two different monomers are added; and condensation, when two different monomers are added, but an inorganic substance is eliminated.

Within soil stabilization studies (Huang et al., 2021), polymers are divided into geopolymers, biopolymers, and synthetic organic polymers. The former are inorganic and are developed from amorphous aluminosilicates activated by concentrated alkaline solutions. Biopolymers are natural polymers produced by biomass or bacteria (Jamshidi et al., 2023; Zinchenko et al., 2022). Finally, synthetic organic polymers are artificial, mainly polyacrylamide (PAM), polyacrylates, polyvinyl acetate, and polyvinyl alcohol.

The use of polymers in geotechnical engineering began to be developed during the Second World War when there was a need to build emergency sidewalks for landing war equipment in coastal regions. Stabilizers such as lime and cement were not feasible due to logistical conditions and the curing process.

Various studies have shown the benefits of using polymers to manage material properties. But, in addition, polymers are more interesting from an environmental point of view because their manufacture requires less consumption of natural sources and energy and also emits less greenhouse gases (GHGs) when compared to traditional stabilizers such as lime and Portland cement (Huang et al., 2021; Nakayenga et al., 2021). Portland cement production is the second most GHG-generating human activity, second only to fossil fuels (Latifi et al., 2016).

The interaction of the polymers with the mineral matrix will depend mainly on the type of polymer used. Generally, the process of aggregating the particles the polymer generates is called flocculation (Prusty et al., 2023). This phenomenon takes place in four stages, as shown in Figure 1. It is important to note that all these actions take place on a microscopic scale, not an engineering scale.

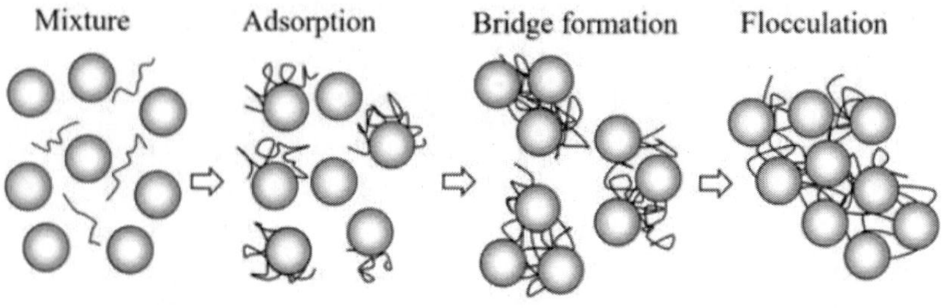

Figure 1. Polymer flocculation process. Developed based on Kitchener (1972).

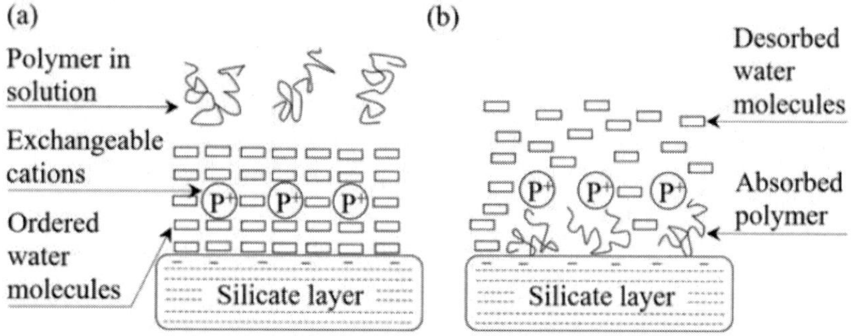

Figure 2. Illustrative diagram depicting the interaction between neutral polymers and clays: (a) Before polymer adsorption; and (b) After desorption of water particles. Developed based on Huang et al., (2021).

Within the group of organic synthetic polymers, there are cationic, anionic, and neutral polymers. In most cases, interaction with the matrix takes place through electrostatic interaction, which is more complex in the case of ionized polymers. For example, neutral polymers interact with clay particles simply by increasing the entropy generated by the combination of two actions: adsorption of the polymer on the surface of the particle concomitant with desorption of water (Prusty et al., 2023), as shown in Figure 2.

On the other hand, cationic polymers work well with clay materials because the surface of these soils is generally negatively charged. This causes the polymer to create bridges on the surface and at the ends of the particle, generating flocculation (Fan et al., 2000).

Anionic polymers, such as the one used in this work, are effective flocculants in cationic environments. In this type, only a few elements of the polymer chain act in adsorption, and most of them are associated with the loops and tails. For this reason, flocculation takes place more easily as the loose structures (loops and tails) create a greater attractive distance between neighboring particles.

On the other hand, anionic polymers tend to be repelled from the surface of the clays and therefore adsorption is low. For this reason, special attention must be paid to the pH of the environment in which they are found. In acidic environments, they can behave like neutral polymers and act between the lamellar particles of the material. In this case, electrostatic interactions are also relevant as they can occur especially with the aluminum exposed on the side of the crystalline structure of minerals, which in acidic conditions acquire positive charges (Fan et al., 2000).

Studies on the interaction of polymers with granular materials, such as sands, have been less explored. This is because the mechanism appears to be simpler since sands generally have neutral charges. Mohamed (2004) believes that the hydraulic and mechanical benefits observed in the interaction of polymers with granular materials are related to three phenomena: (i) covering the particles with a polymeric film; (ii) forming connections between neighboring particles; (iii) developing adhesion between the particles in contact.

Liu (2018) investigated the effect of three types of commercial polymers on the stabilization of a sandy soil using simple compression and permeability tests. For all cases, the authors obtained a significant improvement in strength parameters and a reduction in permeability, especially as the curing period and polymer content increased. Parameter stability was achieved at approximately 14 days.

The effect of the polymer content was also proven by Liu (2018). These authors carried out an unconfined compression test, direct shear test, and splitting tensile test and concluded that in addition to the solution content, the dry specific weight is a determining factor for the strength of the composite. In addition, using optical microscopy tests, the authors observed that the polymers have a cohesive effect between the particles, enveloping them and creating bonds.

Kolay et al., (2016) studied the effect of acrylic polymers on two types of fine soils. To do this, the authors used 5 different contents that were added in relation to the dry specific weight of the soil. The samples were compacted at the optimum point and some of them were cured in contact with air and others in confinement. Unconfined compression tests were carried out for three different curing periods (7, 14, and 28 days). The air-dried samples obtained the best results, with an increase in unconfined compression strength of up to 75%.

Other authors have also followed the same line of experimental program to evaluate soil-polymer composites, such as Mirzababaei et al., (2017), Naeini & Ghorbanali (2010), Okonta (2019), Silva (2020), and Soltani-Jigheh et al. (2019). No research on polymer composites tailings was found in the literature.

Recent Findings and Trends

In an era characterized by rapid technological advancements and a growing awareness of environmental sustainability, the fields of engineering and infrastructure development are constantly evolving to meet the demands of our changing world. This chapter serves as a comprehensive exploration of the latest developments and emerging patterns across three pivotal domains: Mining Site Applications, Transportation Infrastructure Applications, and Landfills Application.

Mining Site Applications

The mining industry, crucial for resource extraction, faces increasing pressure to adapt to environmentally responsible practices. Recent findings in this realm reveal groundbreaking approaches to waste management, safety enhancement, and the optimization of ore production. From the adoption of dry stacking and polymer-based stabilization methods to improve tailings management to the implementation of cutting-edge technologies, this section unveils how the mining sector is striving to strike a balance between economic profitability and ecological preservation.

One noteworthy study, conducted by Carneiro (2020), focused on the mechanical behavior of polymer-stabilized iron ore tailings. This study evaluated the influence of polymer dosage (at proportions of 1:4 and 1:2 in polymer solution), curing time, and load. Furthermore, this investigation also assessed the mechanical behavior of the polymer-tailings composite reinforced with polypropylene fibers (0.5% by weight of dry tailings, randomly distributed), while considering the aforementioned polymer dosages. The morphology of both pure tailings and the composites was evaluated using scanning electron microscopy (SEM), as shown in Figure 3.

Figure 3. SEM analysis results: (a) Pure tailings grains at 110x magnification; (b) Tailings-polymer composite grains at 80x magnification, using a polymer solution ratio of 1:2; and (c) Tailings-polymer-fiber composite grains at 150x magnification, employing a polymer solution ratio of 1:2. Adapted from Carneiro (2020).

In the case of pure tailings (Figure 3a), it is discernible that the grains are loosely arranged and exhibit a heterogeneous texture, with smaller grains filling interstitial spaces between larger ones. Conversely, upon the introduction of the polymer solution (Figure 3b), a noticeable reduction in voids occurs, as the grains visibly approach each other and demonstrate cohesion, implying the efficacy of the curing process in promoting particle bonding. Furthermore, within the framework of the tailings-polymer-fiber composite (Figure 3c), a further reduction in voids is evident, attributable to the ability of the fibers to occupy pre-existing interstices and bring the tailings grains into closer proximity.

As per Carneiro (2020), the stabilization of tailings with polymers yielded favorable outcomes concerning strength parameters, as evidenced by more pronounced peaks in strength for samples subjected to longer curing times, higher confinement stresses, and greater polymer solution content (solution ratio of 1 :2). This resulted in an increase in the values of cohesive intercept and friction angle. A similar behavior was also observed regarding residual stresses, indicating the feasibility of tailings stabilization with polymers.

Based on these results, de Carneiro & Casagrande (2020) emphasize that the application of the waste-polymer composite as a filling material in geotechnical structures, such as landfills and slopes, can contribute significantly to technical, economic, structural, and environmental perspectives.

According to Carneiro (2020), the incorporation of fibers to the tailings-polymer composites exerted a notable influence on the behavior of the polymer-tailings composite, leading to a reduction in stiffness and an enhanced stabilization of residual strength. Importantly, this contributed to the composite's performance even after reaching the strength peak, as the presence of fibers allowed for a greater stress redistribution capacity, enhancing its ability to withstand cyclic loads and minimizing the likelihood of premature failure. This underscores the potential of this composite as a robust and durable material in long-term geotechnical and structural applications.

Another relevant study in the context of mining site applications was conducted by Alelvan et al., (2022) who presented a mechanical and microstructural analysis of pure and polymer-stabilized gold ore tailings, as well as the enhancement of this composite using lime (a field practice for clayey soils stabilized with this polymer) and polypropylene fiber reinforcement. The composites were investigated using a 2k factorial experiment methodology, with two levels established for the factors: polymer content (6% and 9%), curing time (7 and 28 days), and dry unit weight (1.7 g/cm^3 and 1.8 g/cm^3). These materials underwent geotechnical, mechanical, mineralogical, microstructural, and chemical characterization.

This work was conducted to achieve the goals: (1) characterize the pure gold ore tailings, from XRD, XRF, thermogravimetric analysis (TGA), and SEM analysis; (2) verify the effect of polymeric stabilization and fiber reinforcement on the tailing's microstructure, by μCT tests; (3) obtain comparative mechanical responses between the pure tailings and the reinforced and stabilized composites through Unconfined Compression Tests. The research was focused on contributing to the knowledge of tailings composite through microstructural analysis, and its feasibility as construction material by mechanical tests.

The stress-strain curves of the stabilized and reinforced composites presented in Figure 4 show a significant increase in compressive strength compared to the pure tailings. The brittle behavior is common for all curing times and dry unit weight. However, there is an initial stiffness gain with polymer hardening, which leads to smaller axial strains for the composites with longer curing times. Samples at 28 days break with an axial strain of 2% and 3%, while those at 7 days between 3 and 4% for the lower polymer content. For the 9% content, the peak strains at 28 days are 3% and 4.5%, and for 7 days between 4% and 5.5%.

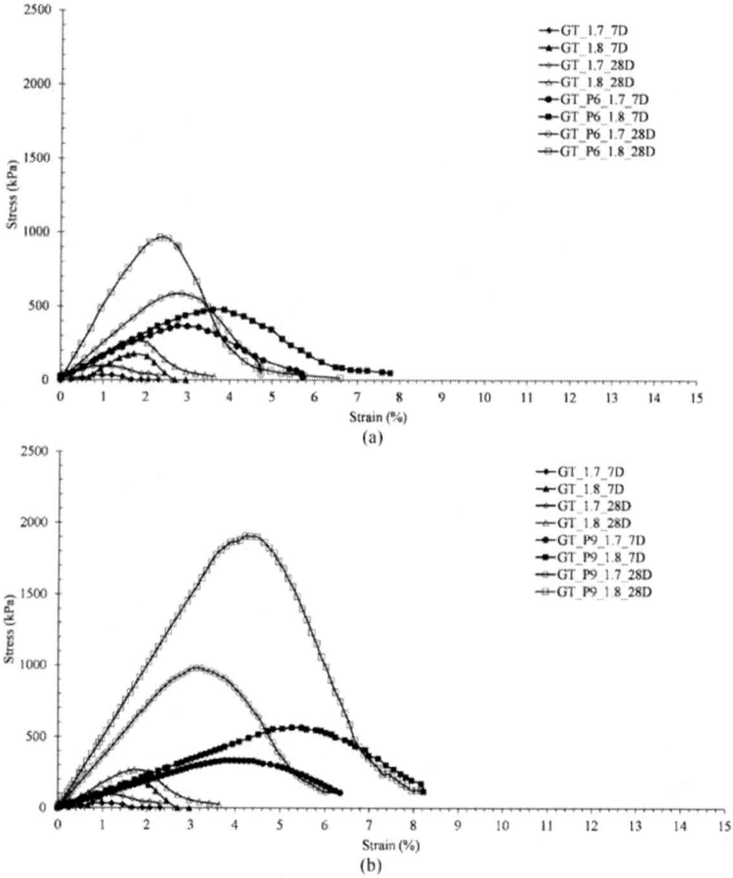

Figure 4. Stress-strain curves for polymer composites. (a) polymer content of 6%; (b) polymer content of 9%. Note: in the graphics caption acronyms are used according to: GT for Gold Tailings, followed by the letter P and the polymer content (X), the value of the dry unit weight (Y.Y), and finally the curing time (ZD), resulting in GT_PX_Y.Y_ZD. (Alelvan et al., 2022).

Figure 5 shows the morphological analysis obtained by SEM. The pure tailings are composed of a lamellar isotextural structure, randomly oriented and with tightly interlocking packets. The tendency to form aggregates in the composite, shown in Figure 5b, is directly related to particle size and surface area availability. However, the lamellar packages present in the pure tailing decrease the surface area for polymer contact, which can lead to brittle behavior. This was an interesting result when compared to the work of de Carneiro & Casagrande (2020), as it shows how the polymer is influenced by the grain size of the matrix.

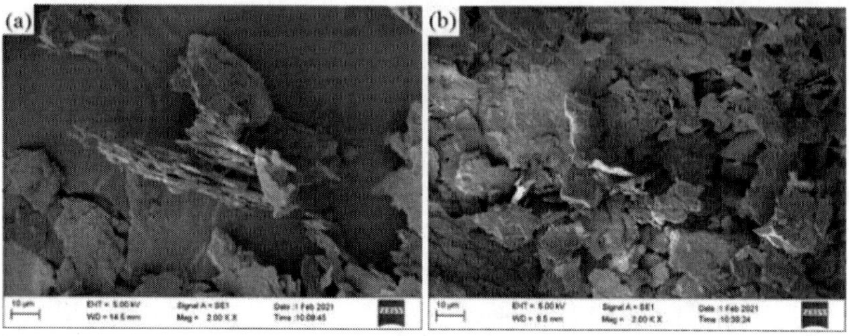

Figure 5. SEM analysis: (a) pure talings; (b) gold ore tailings-polymer composite. (Alelvan et al., 2022).

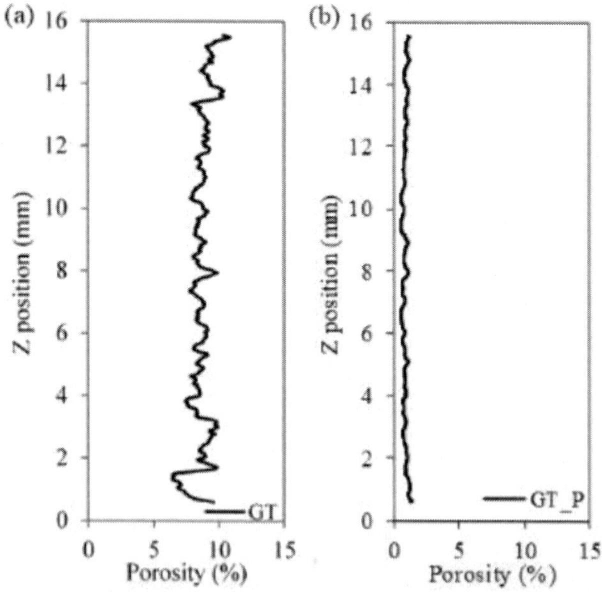

Figure 6. Two-dimensional porosity analysis. (a) pure tailings; (b) gold ore tailings-polymer composite. (Alelvan et al., 2022).

The microtomography analysis enabled the understanding of the behavior of the pure tailings and the composite in several aspects. The two-dimensional porosity analysis was done by evaluating the total, the open, and the closed pores, as Figure 6. The openness of the pores refers to the interconnection between them.

About total porosity, a range of 7 – 10% was obtained for the pure tailings, and much lower values, between 0.5 – 1.5%, for the composite. The same trend was identified by the number of closed pores found in the materials, in the order of 20,000 for the pure tailing, and an average of 7,000 for the composite. This analysis shows that the polymer acts with a cohesive effect by filling the voids in the sample, connecting the particles, and, consequently, decreasing the permeability of the medium.

In the three-dimensional analysis, it is possible to evaluate the volumes of each of the components that constitute the materials. The pure tailings are composed of pores, matrix, clasts, and heavy minerals. In the GT_P composite, besides the elements mentioned, there is the polymer portion. The heavy minerals possess a density greater than 3 g/cm² and are found in the tailings, especially associated with the clasts. The details of each representative volume are presented in Figure 7.

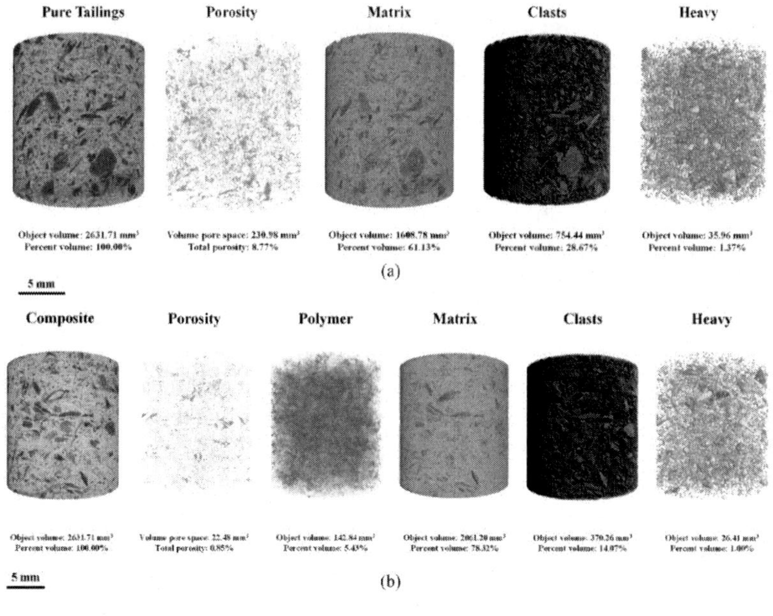

Figure 7. Tridimensal analysis: (a) pure tailings; (b) gold ore tailings-polymer composite. (Alelvan et al., 2022).

The results of the tests conducted by Alelvan (2022) indicate the feasibility of using this stabilizer as an option for reusing gold ore tailings in geotechnical structures. The unconfined compression test of the composite reached values close to 2 MPa, especially for a polymer content of 9% by mass. The incorporation of polymer fibers into the tailings-polymer composite significantly improved the ductile behavior of the material. The results of the direct shear test demonstrated that the polymer's action is in the cohesive effect generated among the particles. Microtomography results confirm this finding, also indicating a decrease in porosity and, consequently, in the permeability of the stabilized material. Finally, it is observed that the management of this solution is simplified in the field, as its application only requires common geotechnical techniques and equipment.

The study conducted by Lopes (2022) addresses the stabilization of ultrafine waste material from iron beneficiation using a polymer solution. The study involved the investigation of the physicochemical, mechanical, mineralogical, and microstructural behavior of composites formed by waste material with varying polymer contents (ranging from 10% to 60% by mass). Incremental increases in polymer dosage corresponded to proportional increases in simple compression strength, enabling the production of samples with higher density. From the perspective of the direct shear test, the addition of polymer resulted in an increase in peak strengths, which shifted to the left, while there were also significant increases in residual strengths. The author also highlights that polymer stabilization led to the agglomeration of waste material grains and the formation of menisci between them, with this behavior linked to particle cementation.

Finally, Sotomayor et al. (2021) address the mechanical behavior of iron and gold ore tailings reinforced with polypropylene fibers, aiming for their applicability in mining waste disposal sites (piles). The fibers were randomly distributed within the tailings (matrix) at a concentration of 0.5% relative to the dry weight of the matrix. The iron ore and gold ore tailings were classified as silty sand and silty clay, respectively. The experimental program included conventional drained and undrained triaxial tests, as well as direct shear tests with a polished surface. The latter was employed to assess residual strength, considering both drained and undrained specimen conditions. Additionally, tests were conducted to evaluate the microstructure through SEM analysis, as indicated in Figure 8.

Based on Figure 8a, it can be observed that, in general, iron tailings grains are larger than the fibers. Sotomayor (2018) emphasizes that 95% of iron tailings grains have a mean diameter greater than that of polypropylene fibers

(0.018 mm), causing the fibers to fill the spaces between the tailing grains, thereby contributing to an increase in the composite's density. On the other hand, following Figure 8b, the author highlights that, in the case of gold tailings, 80% of the fine matrix is smaller than the fiber diameter, leading to the fibers contributing to an increase in matrix porosity. This factor governed the geomechanical behavior of the composites, as discussed subsequently.

Sotomayor (2018) observed that, in the case of composites formed by iron ore tailings fibers, the increase in density provided by the inclusion of fibers favored dilatant behavior during direct shear and triaxial tests, while composites formed by gold tailings fibers exhibited contractive behavior, as the fibers increased matrix porosity. In both cases, there was an enhancement in the shear strength of the tailings with the inclusion of polypropylene fibers. Sotomayor et al., (2021) further emphasize that, for unreinforced tailings, the angle of friction is the dominant parameter for low stresses, and for high stresses, the cohesion parameter becomes more significant. However, in the case of fiber-reinforced tailings, the angle of friction is the dominant parameter in terms of shear strength.

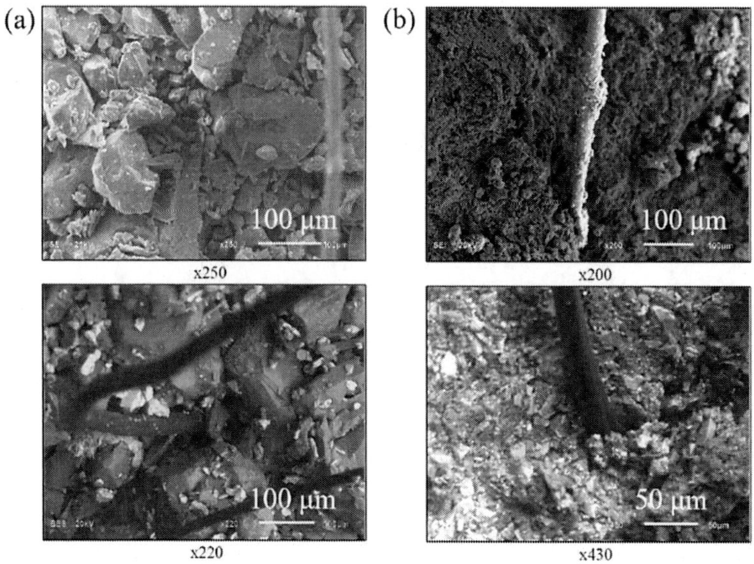

Figure 8. SEM analysis results highlighting the fiber-tailings interaction at different magnifications: (a) Composite composed of iron ore tailings-polypropylene fibers; and (b) Composite encompassing gold ore tailings reinforced with polypropylene fibers Adapted from Sotomayor (2018).

Transportation Infrastructure Applications

As populations grow and urbanization accelerates, the demand for robust transportation networks continues to surge. This section highlights the dynamic trends and discoveries in the domain of Transportation Infrastructure Applications. By addressing the pressing challenges of resilience and sustainability, engineers are forging the path towards more efficient, durable, and environmentally friendly transportation solutions.

It is noteworthy that Carneiro (2020), as mentioned previously, also investigates the application of tailings-polymer and tailings-polymer-fiber composites in transportation infrastructure. The study examines varying polymer dosages (at proportions of 1:4 and 1:2 in polymer solution) and the potential reinforcement with polypropylene fibers (0.5% by weight of dry tailings, randomly distributed). The performance evaluation of these composites in transportation infrastructure applications encompasses tests for diametral compression tensile strength, cyclic triaxial loading (resilience modulus), and wear through a laboratory traffic simulator. In the latter case, the Hamburg Wheel-Track test was employed for traffic simulation, allowing for the assessment of composite wear in response to wheel pass cycles, as depicted in Figure 9.

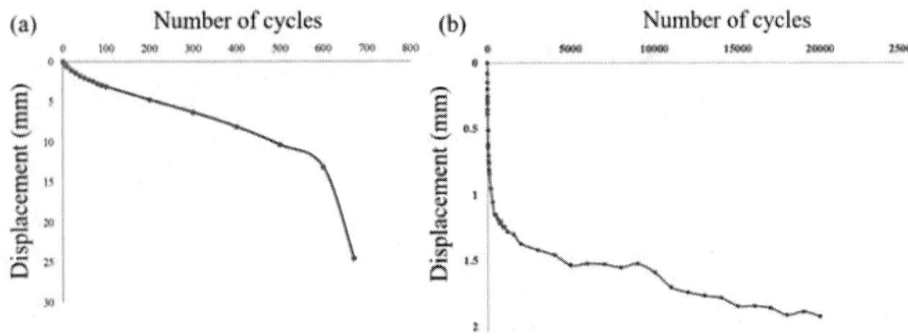

Figure 9. Laboratory traffic simulator wear test results (Hamburg Wheel-Track), illustrating the relationship between the number of cycles and deformation of the composites for two different polymer solution ratios: (a) 1:4; and (b) 1:2. Adapted from Carneiro (2020).

In Figure 9a, the results obtained by Carneiro (2020) for a 1:4 polymer solution ratio are presented, while Figure 9b displays the outcomes corresponding to a 1:2 polymer solution ratio. Based on these findings, it can be inferred that the tailings-polymer composite with a 1:2 polymer solution ratio is suitable for use as a surface layer in pavement applications. Moreover, when evaluating the resilience modulus and comparing it with conventional asphalt mixtures, it was possible to confirm the feasibility of using the 1:2 dosage for unpaved roads with low traffic volume. In the environmental context, leachate analysis did not reveal excessive levels of chemical elements that could pose risks of contaminating soils and groundwater.

Another study investigating the geotechnical performance of a soil-polymer composite was conducted by Silva (2020). It is worth noting that this research did not involve the use of tailings. However, similar to Carneiro (2020), it focused on sandy soil. The soil used in this research originated from an environmental protection area, exhibiting predominantly granular texture and lacking natural cementing agents. This study involved the evaluation of two different polymer dosages (2.5% and 5% relative to the total volume) and various curing times. From a geotechnical perspective, the analysis encompassed physical, mechanical, chemical, environmental, mineralogical, and microstructural aspects. SEM analysis results about the soil in its natural state and after stabilization with polymers are presented in Figure 10.

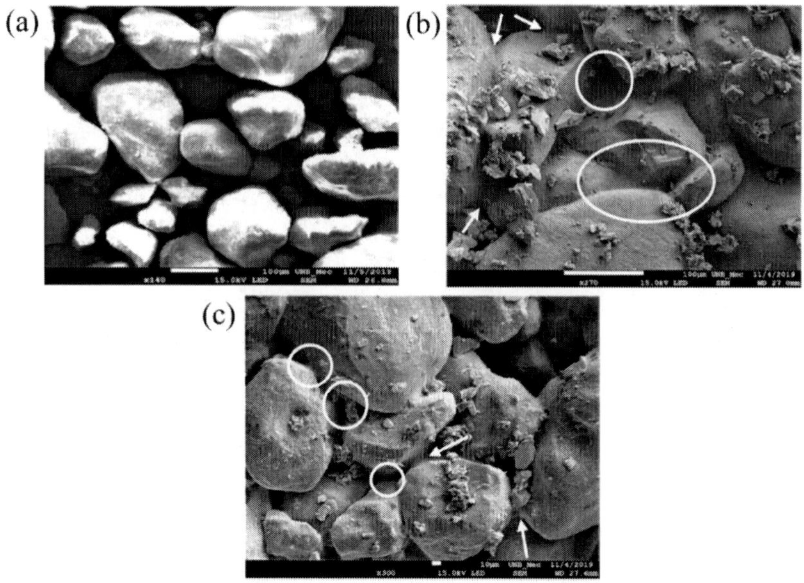

Figure 10. SEM analysis results: (a) Natural state soil at 140x magnification; (b) Soil-polymer composite grains at 270x magnification with a polymer concentration of 2.5%; and (c) Soil-polymer composite grains at 300x magnification with a polymer concentration of 5%. Adapted from Silva (2020).

Based on the SEM images considering the natural state soil (Figure 10a), loose grains with low angularity and heterogeneous granulometry can be observed, with no apparent presence of cementing agents. Conversely, when polymer was added at a concentration of 2.5% (Figure 10b), particle bonding in the substrate can be identified, as indicated by the circles in the figure, while the arrows suggest the presence of menisci uniting the particles and forming small voids. Furthermore, considering a polymer content of 5% (Figure 10c), the presence of particle bonding becomes even more pronounced, imparting an additional cementing component between the particles. Due to this cementation, significant improvements in the strength parameters of the composites were observed for both dosages when compared to untreated soil, with the 5% polymer dosage yielding higher values. There was a progressive increase in cohesive intercept, unconfined compressive strength, resistance to surface deformation, and an elevation in the resilience modulus as higher polymer content in the composites was considered. These results align with those obtained by Lopes (2022).

Regarding applications in transportation infrastructure, Silva (2020) also conducted tests to evaluate expansion, splitting tensile strength, resilience modulus, laboratory traffic simulator (surface wear on DWT equipment), and moisture-induced damage, considering the previously discussed composites. In general, polymer-based stabilization was associated with an increase in the resilient modulus of the composites, with resulting values being compatible with hot mix asphalt. Through chemical analyses and leachate assessments, it was possible to confirm that chemical reactions during the curing process did not lead to an increase in the concentration of contaminating chemical elements, similar to the behavior observed by Carneiro (2020). In the case of applications involving slopes and embankments, the author emphasized the viability of polymer concentrations up to 2.5%. Conversely, for pavement applications, a polymer concentration of 5% was deemed more suitable.

Landfill Applications

The management of waste and land resources is a critical concern in the quest for sustainable urban development. From exploring the integration of green technologies to considering the potential for renewable energy generation from landfill sites, this section underscores the dynamic nature of landfill management and its role in building a more sustainable future.

Traditional methods for waterproofing landfills have historically relied on compacted clay soils to create a low-permeability layer. However, with the increased utilization of geosynthetics, particularly geomembranes and geosynthetic clay liners (GCLs), the risk of subsoil contamination has significantly decreased (Weerasinghe et al., 2021). These modern approaches offer a more effective and reliable means of preventing groundwater pollution compared to relying solely on compacted clay soils.

Geosynthetic Clay Liners (GCLs) are composed of alternating layers of geotextile and high-expansion bentonite. Bentonite, known for its water retention properties, forms a low-permeability layer when hydrated. The geotextiles serve the dual purpose of separation and reinforcement while providing structural support to the bentonite (Maurya et al., 2023).

While Geosynthetic Clay Liners (GCLs) have demonstrated outstanding performance, there remain environmental apprehensions regarding the extensive use of natural clay resources in large-scale applications. In pursuit of a sustainable alternative, researchers at the Federal University of Minas Gerais (UFMG), Brazil, have been investigating the substitution of bentonite in GCLs with mining tailings associated with superabsorbent polymers (SAP) (Oliveira, 2023).

SAP are synthetic materials with a high capacity to absorb water and aqueous solutions that can be more than a thousand times their mass (Bian et al., 2021; Dai et al., 2022). These polymers are made up of long, cross-linked molecular chains that have functional groups capable of binding to water molecules through hydrogen bonds, forming a gel (Friedrich, 2012). SAPs can absorb water up to 5000 times their weight, but in dilute saline solutions, the absorption capacity of commercial products is approximately 50 g/g (Jensen & Hansen, 2001).

The difference between a water-soluble polymer and an absorbent polymer lies in the presence of a crosslinking agent, which is crucial in the water absorption process. This agent is responsible for the formation of crosslinks between the polymer chains and allows water molecules to be trapped and the polymer chain to expand without being dissolved. This reaction allows the formation of a hydrogel that begins to swell when the polymer comes into contact with water, as shown in Figure 11.

Hydrogels expand when they absorb aqueous solutions on their porous surface, retaining the liquid in their structure. This is because hydrogels have cross-links between their main molecular chains and can be explained by osmotic pressure, which causes the water available in the more dilute medium to flow into the polymer network. Superabsorbent polymers are capable of retaining large quantities of liquids without dissolving, and when immersed in water, they expand until they reach their equilibrium volume, thanks to their specific properties. This process can be seen in Figure 12.

To analyze the behavior of this composite material and propose an enhanced composition for Geosynthetic Clay Liners (GCLs), a research initiative at the UFMG was undertaken by Oliveira (2023). The primary focus was on evaluating both the mechanical and hydraulic properties using iron ore tailings. This investigation distinguishes between 'Case 1' and 'Case 2' to elucidate the material characterization and the subsequent experimental campaigns.

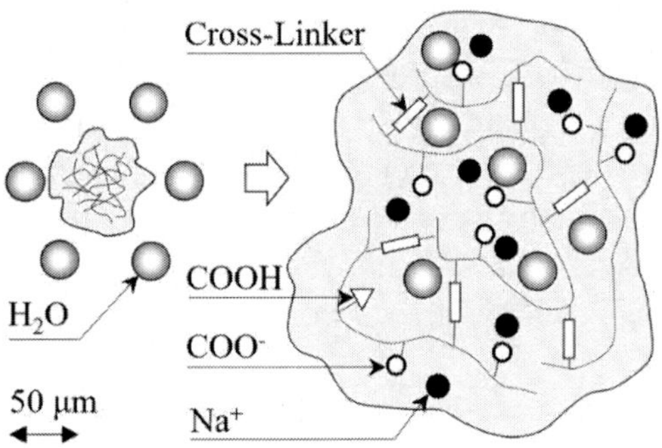

Figure 11. Hydrogel formation includes a representation of a polyacrylate-based SAP particle in both dry and saturated states. Developed based on Friedrich (2012).

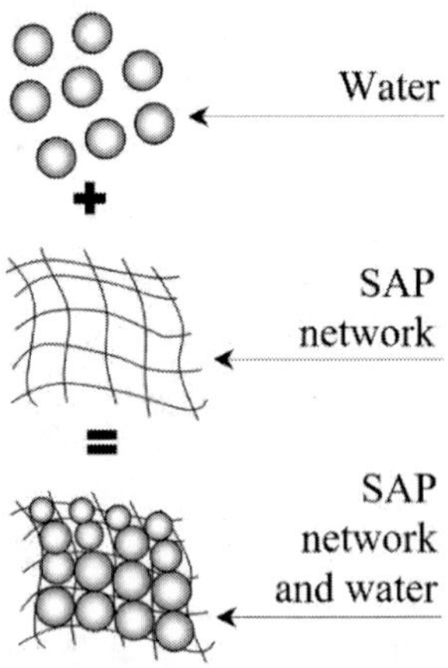

Figure 12. Schematic representation of a 3D SAP network retaining water within its structure Adapted from Manzano et al., (2021).

Case 1 consists of a mechanical assessment of the behavior of an iron ore tailing stabilized with a sodium-based SAP. To this end, tests were carried out with dosages of 1% (SP_1%) and 3% (SP_3%) polymer in relation to the mass of tailings. The composites were tested in simple compression for 7 and 28 days of curing. This tailing was classified as low-compressibility silt according to the Unified Soil Classification System (USCS). The Standard Proctor Compaction test showed that the inclusion of SAP altered the optimum moisture content and dry density, as shown in Table 1.

Unconfined compressive strength tests were systematically conducted on the composite specimens, and the resultant data is illustrated in Figure 13. The composites were molded for two moisture contents: 28% and 35%. The first value refers to the optimum content for the composite of 1%. The pure tailings were molded at their optimum content of 15%, to evaluate their maximum performance. All tests were carried out in triplicate.

Table 1. Standard proctor compaction results

Sample	Optimum moisture content (%)	Dry density (g/cm³)
PT	15.5	2.06
SP_1%	28.0	1.91
SP_3%	26.0	1.60

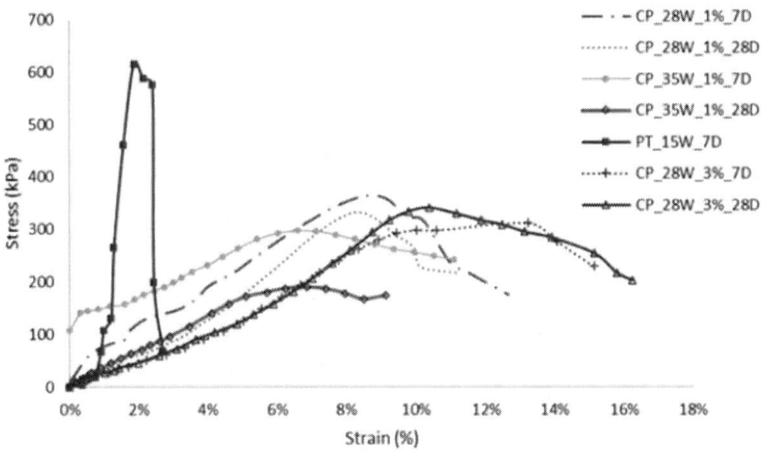

Figure 13. Unconfined compression test results for Case 1. Note: in the graphics caption acronyms are used according to: CP for Composites, followed by the letter moisture content value (XW), polymer content percentage (Y%), and finally the curing time (ZD), resulting in GP_XW_Y%_ZD.

A consistent trend was observed, demonstrating a reduction in strength and an augmentation in deformability corresponding to an increased polymer content within the composites. Notably, specimens with an excess of water (moisture content of 35%) showcased unfavorable tendencies. The loss of strength was distinctly pronounced, while the increase in deformability was comparatively subdued in contrast to other specimens. Consequently, the optimal moisture content is identified as the requisite condition for attaining peak strength within the composite structure.

The outcomes unequivocally establish that the polymer component does not manifest a substantial influence on strength augmentation. However, it significantly enhances deformability properties, substantiating an elastoplastic behavior for the polymer within the composite matrix. The scrutiny of the pure tailings test curve, as well as the compression process, unveiled the inherent brittle behavior distinctive to tailing materials.

Figure 14 shows the specimens after the simple compression test. In addition to the fragile behavior of the pure tailings, it can be seen that the composite has a porous appearance. The voids seen in the image are the hydrogels formed by the SAP which, after evaporation, leave the pores. For this reason, no improvement in mechanical behavior can be expected in non-saturated conditions.

Figure 14. Unconfined compression test samples: pure tailings and composites (tailings + SAP).

Table 2. Permeability test results for Case 2

Sample	Permeability coefficient (cm/s)
PT2	$7.86 \cdot 10^{-3}$
SP_0.05%	$3.80 \cdot 10^{-3}$
SP_0.25%	$5.23 \cdot 10^{-4}$
SP_0.75%	$1.80 \cdot 10^{-5}$

Case 2 consists of analyzing the hydraulic behavior of mining tailings with SAP using constant load permeability tests. The tailings in this case consist of a granular material, classified as poorly graded sand. The test was carried out for pure tailings (PT2) and three polymer contents: 0.05%, 0.25%, and 0.75%. The results are shown in Table 2.

One concern with SAP is its resorption capacity. In this same study, the researchers evaluated three cycles of the composite, starting with oven drying at 110°C for at least 15 days. The results showed that there was no loss in resorption capacity, leading to the same permeability coefficient values.

Much remains to be thoroughly assessed concerning the integration of Superabsorbent Polymers (SAP) in geotechnical applications. Nevertheless, the obtained results hold promise for utilizing this material as an effective waterproofing solution. The synergistic integration of SAP with mining waste presents an additional advantage, particularly in the context of sustainable resource management for civil construction purposes.

Final Considerations

This chapter highlights the mining industry's commitment to addressing environmental challenges while maintaining economic viability. The transition from traditional tailings dams to innovative solutions benefiting from polymer stabilization and polypropylene fiber reinforcements offers a promising direction. The comprehensive assessment of the physical, chemical, mechanical, environmental, and microstructural characteristics of different solutions has revealed the transformative potential of these methods. Based on the information presented in this paper, the following conclusions may be drawn:

- Carneiro (2020) and de Carneiro & Casagrande (2020) investigations demonstrated that polymer-stabilized iron ore tailings, particularly those with higher polymer content and extended curing times, displayed enhanced strength parameters, suggesting the potential for tailings stabilization using polymers. Furthermore, Carneiro's study in 2020 examined the utilization of tailings-polymer composites in transportation applications, affirming their suitability for both unpaved road construction and pavement surface layers.

- Silva (2020) study investigated two polymer dosages (2.5% and 5%) and various geotechnical aspects of sand-polymer composites. SEM analysis revealed enhanced particle bonding and cementation in the stabilized composites, leading to improved strength parameters, with the 5% polymer dosage showing higher values. The findings suggested that polymer concentrations of up to 2.5% are suitable for slopes and embankments, while for pavement applications, a 5% polymer concentration is recommended.
- The investigations by Sotomayor (2018) and Sotomayor et al., (2021) explore polypropylene fiber reinforcement in iron and gold ore tailings for mining waste disposal. Their experiments reveal improved shear strength in both tailing types, with iron ore tailings showing dilatant behavior due to increased density and gold ore tailings displaying contractive behavior due to higher matrix porosity.
- The studies by Alelvan (2022) and Alelvan et al., (2022) evaluate the mechanical and microstructural properties of gold ore tailings, both in their pure form and when stabilized with a polymer and reinforced with polypropylene fibers. The findings reveal that the addition of the polymer and fibers significantly enhances the strength and ductility of the composite material, making it a promising option for reuse in geotechnical structures. Microtomography results further support this, showing reduced porosity and increased cohesion.
- Superabsorbent polymers (SAP), known for their ability to absorb substantial amounts of water and aqueous solutions, hold promise for a wide range of applications. These hydrogels, created using SAP, expand as they absorb liquids, demonstrating the potential

for moisture management and offering environmentally friendly solutions in landfill management and other fields.

As the mining sector adopts sustainability and resilience principles, these discoveries emphasize the significance of polymers in improving tailings, broadening their applicability in engineering projects such as tailings storage, embankments, and road surfacing. It is important to note that while the discussions presented in this chapter contribute to the ongoing conversation about responsible resource management and engineering excellence in mining, a more in-depth assessment of the life cycle of these alternative solutions should be a key focus of future research. This aspect represents one of the most relevant areas of investigation in the near future.

As examples of prospects for the application of new geomaterials related to mining activities, it is worth highlighting the use of nanoparticles and nanopolymers, which have shown potential for use in geotechnical works. Nevertheless, further in-depth investigations regarding their behavior are required. Finally, the authors would like to emphasize that the term "stabilization" needs to be better evaluated, as the interfacial phenomena involved vary in each type of application. This necessitates the continuous development of research addressing the use of new geotechnical materials.

References

Alelvan, G. M. (2022). *Análise mecânica e microestrutural de rejeito de minério de ouro estabilizado com solução polimérica*. University of Brasília.

Alelvan, G. M., Ferreira, J. W. D. S., Casagrande, M. D. T. & Consoli, N. C. (2022). Proposal of new construction material: Polymer-stabilized gold ore tailings composite. *Sustainability*, *14*(20), 13648. https://doi.org/10.3390/su142013648.

Alshawmar, F. & Fall, M. (2023). Investigation of drying and wetting effects on response of highly densified tailings to cyclic loadings: Shaking table test results. *Soil Dynamics and Earthquake Engineering*, *166*, 107769. https://doi.org/10.1016/j.soildyn.2023.107769.

Bian, X., Zeng, L., Li, X., Shi, X., Zhou, S. & Li, F. (2021). Fabric changes induced by super-absorbent polymer on cement–lime stabilized excavated clayey soil. *Journal of Rock Mechanics and Geotechnical Engineering*, S1674775521000470. https://doi.org/10.1016/j.jrmge.2021.03.006.

Cao, L., Zhou, J., Zhou, T., Dong, Z. & Tian, Z. (2022). Utilization of iron tailings as aggregates in paving asphalt mixture: A sustainable and eco-friendly solution for mining waste. *Journal of Cleaner Production*, *375*, 134126. https://doi.org/10.1016/j.jclepro.2022.134126.

Carneiro, A. A. (2020). *Comportamento mecânico de um rejeito de minério de ferro estabilizado com polímero e do compósito rejeito-polímero reforçado com fibras de polipropileno*. University of Brasília.

Carneiro, A. A. & Casagrande, M. D. T. (2020). Mechanical and environmental performance of polymer stabilized iron ore tailings. *Soils and Rocks*, *43*(4), Artigo 4. https://doi.org/10.28927/SR.434679.

Castillo, C., Fawell, P. & Costine, A. (2023). Optimising the activity of acrylamide-based polymer solutions used to flocculate mineral processing tailings suspensions – A review. *Chemical Engineering Research and Design*, *199*, 214–237. https://doi.org/10.1016/j.cherd.2023.10.001.

Cota, F., Corrêa, A., Viana, Í. & Visacro, S. (2022). Modeling soil ionization effect based on experimental data. *Electric Power Systems Research*, *213*, 108762. https://doi.org/10.1016/j.epsr.2022.108762.

Dai, D., Peng, J., Wei, R., Li, L. & Lin, H. (2022). Improvement in dynamic behaviors of cement-stabilized soil by super-absorbent-polymer under cyclic loading. *Soil Dynamics and Earthquake Engineering*, *163*, 107554. https://doi.org/10.1016/j.soildyn.2022.107554.

Fan, A., Turro, N. J. & Somasundaran, P. (2000). A study of dual polymer flocculation. *Colloids and Surfaces A: Physicochemical and Engineering Aspects*, *162*(1–3), 141–148. https://doi.org/10.1016/S0927-7757(99)00252-6.

Friedrich, S. (2012). Superabsorbent Polymers (SAP). Em V. Mechtcherine & H.-W. Reinhardt (Orgs.), *Application of super absorbent polymers (SAP) in concrete construction* (p. 13–19). Springer Netherlands. https://doi.org/10.1007/978-94-007-2733-5_3.

Galjak, J., Đokić, J., Milentijević, G., Dervišević, I. & Jović, S. (2020). Characterization of the tailing waste deposit "Gornje Polje". *Optik*, *215*, 164684. https://doi.org/10.1016/j.ijleo.2020.164684.

Gao, Z., Li, Y., Qian, H. & Wei, M. (2023). Environmental, economic, and social sustainability assessment: A case of using contaminated tailings stabilized by waste-based geopolymer as road base. *Science of The Total Environment*, *888*, 164092. https://doi.org/10.1016/j.scitotenv.2023.164092.

Huang, J., Kogbara, R. B., Hariharan, N., Masad, E. A. & Little, D. N. (2021). A state-of-the-art review of polymers used in soil stabilization. *Construction and Building Materials*, *305*, 124685. https://doi.org/10.1016/j.conbuildmat.2021.124685.

Jamshidi, M., Mokhberi, M., Hossein Vakili, A. & Nasehi, A. (2023). Effect of chitosan bio-polymer stabilization on the mechanical and dynamic characteristics of marl soils. *Transportation Geotechnics*, *42*, 101110. https://doi.org/10.1016/j.trgeo.2023.101110.

Jensen, O. M. & Hansen, P. F. (2001). Water-entrained cement-based materials. *Cement and Concrete Research*, *31*(4), 647–654. https://doi.org/10.1016/S0008-8846(01)00463-X.

Jiang, P., Wang, Z., Wang, W., Li, N., Yu, Y., Li, C., Wu, E. & Pu, S. (2023). Interfacial strength characteristics between modified iron tailings and profiled fibers under dry–

wet and freeze–thaw environments. *Construction and Building Materials*, *403*, 133016. https://doi.org/10.1016/j.conbuildmat.2023.133016.

Khodabandeh, M. A., Nagy, G. & Török, Á. (2023). Stabilization of collapsible soils with nanomaterials, fibers, polymers, industrial waste, and microbes: Current trends. *Construction and Building Materials*, *368*, 130463. https://doi.org/10.1016/j.conbuildmat.2023.130463.

Kitchener, J. A. (1972). Principles of action of polymeric flocculants. *British Polymer Journal*, *4*(3), 217–229. https://doi.org/10.1002/pi.4980040307.

Kolay, P. K., Dhakal, B., Kumar, S. & Puri, V. K. (2016). Effect of liquid acrylic polymer on geotechnical properties of fine-grained soils. *International Journal of Geosynthetics and Ground Engineering*, *2*(4), 29. https://doi.org/10.1007/s40891-016-0071-5.

Krishna, R. S., Quezada, G. R., Sahu, J. K. & Sadangi, J. K. (2021). Rheological characterization and performance of flocculants in iron ore tailings management. *Materials Today: Proceedings*, *43*, 2888–2894. https://doi.org/10.1016/j.matpr.2021.01.128.

Lambe, T. W., & Whitman, R. V. (1969). *Soil mechanics*. Wiley.

Latifi, N., Rashid, A. S. A., Siddiqua, S. & Majid, Muhd. Z. A. (2016). Strength measurement and textural characteristics of tropical residual soil stabilised with liquid polymer. *Measurement*, *91*, 46–54. https://doi.org/10.1016/j.measurement.2016.05.029.

Liu, X. (2018). Response of sandy soil stabilized by polymer additives. *Open Access Journal of Environmental and Soil Sciences*, *1*(3). https://doi.org/10.32474/OAJESS.2018.01.000112.

Lopes, M. M. (2022). *Análise mecânica e microestrutural de um rejeito de minério de ferro ultrafino estabilizado com solução polimérica*. University of Brasília.

Manz, T. A. (2024). Density-derived electrostatic and chemical methods. Em *Comprehensive Computational Chemistry* (p. 362–405). Elsevier. https://doi.org/10.1016/B978-0-12-821978-2.00072-6.

Manzano, M. A. R., Silva, E. F. D., Lopes, A. N. D. M. & Tolêdo Filho, R. D. (2021). Mecanismo de atuação dos polímeros superabsorventes como agentes de cura interna para mitigar a retração autógena em concretos de alta resistência (CAR) – Estado da Arte. *Matéria (Rio de Janeiro,* in Portuguese*)*, *26*(2), e12956. https://doi.org/10.1590/s1517-707620210002.1256.

Maurya, S. D., Dixit, A. & Singh, D. (2023). Application of peanut shell ash in geosynthetic clay liners to reduce the zinc contamination in leachate. *Materials Today: Proceedings*, S2214785323031723. https://doi.org/10.1016/j.matpr.2023.05.536.

Mirzababaei, M., Arulrajah, A. & Ouston, M. (2017). Polymers for stabilization of soft clay soils. *Procedia Engineering*, *189*, 25–32. https://doi.org/10.1016/j.proeng.2017.05.005.

Mohamed, S. W. A. (2004). *Stabilization of Desert Sand Using Water-Borne Polymers*. United Arab Emirates University.

Naeini, S. A., & Ghorbanali, M. (2010). Effect of wet and dry conditions on strength of silty sand soils stabilized with epoxy resin polymer. *Journal of Applied Sciences*, *10*(22), 2839–2846. https://doi.org/10.3923/jas.2010.2839.2846.

Nakayenga, J., Cikmit, A. A., Tsuchida, T. & Hata, T. (2021). Influence of stone powder content and particle size on the strength of cement-treated clay. *Construction and Building Materials*, *305*, 124710. https://doi.org/10.1016/j.conbuildmat.2021.124710.

Nekhoul, B., Harrat, B., Boutadjine, A. & Melit, M. (2022). A simplified numerical modeling of the transient behavior of grounding systems considering soil ionization. *Electric Power Systems Research*, *211*, 108182. https://doi.org/10.1016/j.epsr.2022.108182.

Okewale, I. A. & Grobler, H. (2023). Assessment of heavy metals in tailings and their implications on human health. *Geosystems and Geoenvironment*, *2*(4), 100203. https://doi.org/10.1016/j.geogeo.2023.100203.

Okonta, F. (2019). Pavement geotechnical properties of polymer modified weathered semi-arid shale subgrade. *International Journal of Pavement Research and Technology*, *12*(1), 54–63. https://doi.org/10.1007/s42947-019-0007-2.

Oliveira, G. A. (2023). *Estabilização Química de Rejeito de Minério de Ferro com Polímero Superabsorvente*. Federal University of Minas Gerais.

Oliveira Junior, G. G. de. (2011). *Desenvolvimento de uma rota de processo de dessulfurização de rejeitos de um minério aurífero sulfetado*. Federal University of Minas Gerais.

Pan, M., Wang, W., Duan, C., Jiang, H., Zhao, Y., Qiao, J., Huang, L., Wang, Z., Shen, Y. & Shi, W. (2022). Process enhancement of vibrating classifier for tailings classification-dewatering and industrial application. *Powder Technology*, *400*, 117219. https://doi.org/10.1016/j.powtec.2022.117219.

Pierozan, R. C., Araújo, G. L. S., Palmeira, E. M. & Romanel, C. (2023). Influence of variables related to soil weathering on the geomechanical performance of tropical soils. *Journal of Rock Mechanics and Geotechnical Engineering*, *15*(9), 2423–2440. https://doi.org/10.1016/j.jrmge.2023.04.005.

Prusty, D., Gallegos, A. & Wu, J. (2023). Adsorption behavior of associating nanoparticle-polymer systems in the vicinity of an attractive surface: Predictions from classical density functional theory. *Journal of Molecular Liquids*, *390*, 122976. https://doi.org/10.1016/j.molliq.2023.122976.

Silva, N. A. B. S. (2020). *Desempenho de um compósito solo-polímero para aplicabilidade em obras geotécnicas e de pavimentação.* University of Brasília.

Smuda, J., Dold, B., Spangenberg, J. E., Friese, K., Kobek, M. R., Bustos, C. A. & Pfeifer, H.-R. (2014). Element cycling during the transition from alkaline to acidic environment in an active porphyry copper tailings impoundment, Chuquicamata, Chile. *Journal of Geochemical Exploration*, *140*, 23–40. https://doi.org/10.1016/j.gexplo.2014.01.013.

Soltani-Jigheh, H., Bagheri, M. & Amani-Ghadim, A. R. (2019). Use of hydrophilic polymeric stabilizer to improve strength and durability of fine-grained soils. *Cold Regions Science and Technology*, *157*, 187–195. https://doi.org/10.1016/j.coldregions.2018.10.011.

Sotomayor, J. M. G. (2018). *Avaliação do comportamneto mecânico drenado e não renado de rejeito de minérios de ferro e de ouro reforçados com fibras de polipropileno.* Pontifical Catholic University of Rio de Janeiro.

Sotomayor, J. M. G., Alelvan, G. M. & Casagrande, M. D. T. (2021). Influence of Polypropylene Fiber-Reinforcement on the Mechanical Behavior of Gold Ore Tailings through Direct Shear Tests. *Journal of Materials in Civil Engineering*, *33*(10), 04021256. https://doi.org/10.1061/(ASCE)MT.1943-5533.0003878.

Wayal, V., Sitharam, T. G. & Anjali, M. (2021). Geo-electrical characterization of physical and mechanical properties of zinc tailing. *Journal of Applied Geophysics*, *188*, 104315. https://doi.org/10.1016/j.jappgeo.2021.104315.

Weerasinghe, I. A., Gallage, C. & Dawes, L. (2021). Effect of overburden confining stress on hydraulic performance of geosynthetic clay liners (GCLs). *Heliyon*, *7*(1), e05770. https://doi.org/10.1016/j.heliyon.2020.e05770.

Zhang, A., Bain, J. G., Schmall, A., Ptacek, C. J. & Blowes, D. W. (2023). Geochemistry and mineralogy of legacy tailings under a composite cover. *Applied Geochemistry*, 105819. https://doi.org/10.1016/j.apgeochem.2023.105819.

Zimbardo, M., Ercoli, L., Mistretta, M. C., Scaffaro, R. & Megna, B. (2020). Collapsible intact soil stabilization using the non-aqueous polymeric vehicle. *Engineering Geology*, *264*, 105334. https://doi.org/10.1016/j.enggeo.2019.105334.

Zinchenko, A., Sakai, T., Morikawa, K. & Nakano, M. (2022). Efficient stabilization of soil, sand, and clay by a polymer network of biomass-derived chitosan and carboxymethyl cellulose. *Journal of Environmental Chemical Engineering*, *10*(1), 107084. https://doi.org/10.1016/j.jece.2021.107084.

Chapter 3

Sustainable Solutions and Methods for Swelling Soil Stabilization

Carina Silvani[1,*]
João Pedro Camelo Guedes[2]
Eduardo Antonio Guimarães Tenório[3]
and Samuel Gadelha Guimarães[4]

[1]Graduate Program in Civil and Environmental Engineering, Universidade Federal de Campina Grande, Campina Grande, PB, Brazil
[2]PhD Candidate in Graduate Program in Civil Engineering, Universidade Federal do Rio Grande do Sul, Porto Alegre, RS, Brazil
[3]Undergraduate Program in Civil Engineering, Instituto Federal do Sertão Pernambucano, Serra Talhada, PE, Brazil
[4]Undergraduate Program in Civil Engineering, Instituto Federal de Pernambuco, Afogados da Ingazeira, PE, Brazil

Abstract

This research investigates sustainable solutions for stabilizing swelling soils within the context of environmental geotechnics, waste management, and geotechnical engineering. Swelling soils pose significant issues in civil infrastructure due to their volumetric changes in response to moisture content fluctuations. Although lime and cement have been traditionally used for chemical stabilization to mitigate swelling soil problems, they raise environmental concerns regarding carbon dioxide emissions, energy consumption, and depletion of non-renewable resources. To address these issues, this study explores the utilization of industrial and organic by-products, including rice husk ash

* Corresponding Author's Email: carinasilvani@gmail.com.

In: Environmental Geotechnics
Editors: E. Pavan Korf, P. Domingos Marques Prietto, C. Müller et al.
ISBN: 979-8-89113-651-9

(RHA), rock dust waste (RDW), sugarcane bagasse ash (SCBA), fly ash (FA), brackish water (BW), polyethylene terephthalate (PET), and electric arc furnace (EAF) slag. These materials, whether used alone or in combination with chemical additives, demonstrate potential as effective stabilizing agents. Their implementation not only reduces waste generation but also promotes sustainability in geotechnical projects. Experimental investigations primarily assessed the mechanical behavior and performance of the stabilized materials through freeze-thaw or wet-dry cycles for durability, one-dimensional swelling tests for swell analysis, and unconfined compressive strength (UCS) tests. Additionally, chemical analyses employing Scanning Electron Microscopy (SEM), methylene blue index tests, and x-ray diffraction (XRD) were conducted to evaluate the influence of additives on the mineralogical responses of swelling soils. The results highlight that the incorporation of industrial by-products additives provides efficient and feasible alternatives for stabilizing swelling soils. Certain materials, such as SCBA, FA, BW, EAF, and RHA, contribute to soil stabilization through chemical reactions, whereas RDW and PET substitute the expansive material with more stable alternatives. These additives not only enhance the engineering properties and long-term stability of swelling soils but also promote environmental sustainability by utilizing locally available resources and reducing reliance on energy-intensive materials.

Keywords: expansive soils, chemical stabilization, physical stabilization, waste management, eco-friendly materials

Swelling Soils

Expansive soils present a significant global challenge within the field of geotechnical engineering. These soils, which are characterized as non-saturated, plastic clay soils, undergo substantial volume changes in response to variations in moisture content. As moisture levels increase, they exhibit considerable expansion, and conversely, they shrink as moisture levels decrease (Lambe & Whitman, 1959; Puppala et al. 2013).

During dry periods, these soils develop extensive surface cracking due to soil shrinkage. Changes in the properties of expansive clay soils are associated with various phenomena, including a reduction in shear strength as they absorb moisture, resulting in negative pore pressures. Additionally, alterations in the soil's macrostructure occur due to cyclic processes induced by climatic conditions. According to Snethen (1979), semi-arid climates, characterized by significant wetting and drying cycles, induce detrimental behavior in expansive soils, leading to cyclic expansion and shrinkage each year. Over several years or cycles, this process results in subsoil cracking, leading to substantial differential movements in various types of constructions (Consoli et al. 2021; Phanikumar & Singla, 2016).

These soils can experience volumetric expansion of up to 30% from their initial state as the moisture content increases. These volumetric variations can trigger numerous issues and damage to buildings and place significant stress on civil infrastructure, including buildings, roads, and railways. Foundations constructed on expansive soils are susceptible to substantial uplift forces, leading to cracks, ruptures, and other structural issues. Conversely, soil shrinkage due to moisture loss can compromise foundation support, leading to differential settlement, subsidence, and increased water infiltration into deeper layers, further aggravating soil cracking. This cyclic volumetric variation phenomenon induces fatigue in the soil, making it more vulnerable to erosion, which escalates to various structural problems such as pavement cracks, subsoil cracking, and failures in structural elements (Barman & Dash, 2022; Ikeagwuani et al. 2019).

Data indicates that around the 1970s, the United States allocated approximately 2 billion dollars annually to geotechnical solutions addressing expansive soil-related issues. In more recent periods, China and the United Kingdom have been spending millions of dollars annually on soil and structure maintenance and repair, while Australia has identified expansive soil-related problems in approximately 20% of its national territory (Consoli et al. 2019; Ferreira et al. 2018; Kulanthaivel et al. 2021).

Clay Minerals Composition

Puppala et al. (2013) have noted that the volumetric instability observed in expansive soils can be attributed to certain internal factors, including geological formation, mineralogical and chemical composition, the laminar structure of clay minerals, density, and moisture content. These clay minerals originate from the weathering of rocks, breaking down into smaller soil particles that eventually form clay components. The clay minerals can take on different arrangements of multiple layers, which can manifest in either a 1:1 ratio (kaolinite) or a 2:1 ratio (smectite). Particularly, among these smectite clay minerals, montmorillonite is noteworthy.

Montmorillonite and similar clay minerals possess a 2:1 structure, characterized by the molecular arrangement of two tetrahedral layers (silicon layer) and one octahedral layer (aluminium layer). This structure inherently presents natural imperfections due to chemical charge imbalances. Clay minerals featuring the 2:1 structure exhibit significant expansive properties, primarily because of the weak intermolecular attraction forces between adjacent unit cells. These weak forces result in a more considerable interlayer spacing when compared to other structures, such as the clay mineral with 1:1 proportion (Mitchell & Soga, 2005).

The structural charge deficiency and the presence of a weaker Van der Waals bond between successive silicon layers facilitate the rapid entry of water and exchangeable cations, leading to the separation of primary layers. Consequently, soils rich in montmorillonite are particularly susceptible to expansive behavior (Phanikumar & Singla, 2016; Puppala et al. 2013).

Soil Expansion: Mechanisms and Implications

Clay particles are known to possess negative surface charges due to isomorphic substitution. This results in electrostatic forces between the negatively charged clay surface and the exchangeable cations found in the pore fluid of the clay. The strength of these forces depends on the chemistry of the exchangeable cation, including its charge, position, and valency. To maintain electrical neutrality, ions naturally are attracted to the surface of clay particles. Consequently, an "electric diffuse double layer" forms around these clay particles (Olphen, 1977).

The quantity of cations needed to balance the negative surface charge of the clay mineral is referred to as Cation Exchange Capacity (CEC). A soil's potential for expansion is linked to the structural arrangement of its clay minerals, CEC, and soil suction. Soils with a higher specific surface area typically exhibit a greater CEC, increased surface activity, and consequently, elevated water retention capacity (Mitchell & Soga, 2005).

Generally, when water infiltrates the soil, it hydrates the exchangeable cations located between the internal surfaces of clay minerals, reducing the attractive forces between the layers. This allows more water to penetrate between the clay mineral layers, resulting in additional forces and an expansion of the interlayer spacing. Furthermore, a diffuse double layer forms between the particles, generating an electrical repulsive force due to electrostatic interactions involving the negative surface charge, ions within the clay, and water molecules in the voids. This contributes to further expansion (Ikeagwuani et al. 2019; Nalbantoğlu, 2004).

Among the various ions available to neutralize the negative charge of the clay mineral structures, Ca^{2+}, Mg^{2+}, Na^{+} and H^{+} are the most common, although others like K^{+}, Al^{3+}, Fe^{2+}, Fe^{3+} e Li^{+} occur less frequently (Seco et al. 2011).

When water is introduced, the chemical and hydration properties of Na^{+} cause the crystals to disperse and separate. This creates an electrical potential that results in repulsion between particles. It is worth noting that cations with higher valence easily replace those with lower valence. Therefore, active clays contain a significant concentration of sodium cations available for ion exchange, whereas less active clays typically contain divalent cations with a lower affinity for water (Bolt, 2015; Mitchell & Soga, 2005).

Swelling Soil Treatment

Effective management of expansive soils is crucial to mitigate their detrimental effects. Various procedures and techniques can be employed to address these issues, and chemical soil stabilization proves to be a viable and effective solution in many cases. This method involves adding materials to expansive soil, such as lime, cement, fly ash, rock dust, chlorides, and other alternative waste materials. Chemical stabilization utilizes additives to control soil swelling through processes like soil flocculation and pozzolanic reactions (Consoli et al. 2021; Nalbantoğlu, 2004; Silvani et al. 2020; Tiwari et al. 2021).

The application of these additives, such as cement, lime, and ashes, increases electrical conductivity by raising the concentration of multivalent cations in the pore water (e.g., Ca^{+2} and OH^{-}). This leads to a cation exchange with monovalent ions (e.g., Na^{+}) found in the diffuse double layer of clay particles. This exchange results in an increase in osmotic potential and a decrease in repulsive forces between clay surfaces, consequently, reducing the soil's expansion capacity. Simultaneously, cation exchange between clay and the stabilizing agent, combined with capillary tension between clay particles, promotes flocculation and reduces available surfaces for interaction with water, further decreasing the soil's expansive potential (Dang et al. 2021; Mitchell & Soga, 2005).

Additionally, pozzolanic reactions occur between the calcium ions from the additive and the silica and alumina minerals of clay minerals. These reactions result in the formation of cementitious products such as calcium silicate hydrates (C-S-H), calcium aluminate hydrates (C-A-H), and calcium aluminum silicate hydrates (C-A-S-H) (Mitchell & Soga, 2005).

These pozzolanic products contribute to the flocculation process, binding adjacent flocculated soil particles and thus reducing the soil's expansion potential. As the curing time extends, soil strength increases due to the continuous formation of new cementing phases. These phases interconnect clay clusters and increasingly fill the void spaces resulting from short-term reactions. This continuous bonding reduces clay permeability, leading to a decrease in hydration and, consequently, a reduction in soil expansion capacity (Seco et al. 2011).

Traditional Solution for Swelling Soil

Lime

Lime is a binder produced by the calcination of limestone or dolomite through thermal decomposition and subsequent hydration to form calcium hydroxide, $Ca(OH)_2$. It is important to note that lime does not inherently contain pozzolanic materials. However, when added to the soil alongside a pozzolan source, it triggers reactions such as pH elevation, cation exchange, flocculation, particle agglomeration (due to the influence of Ca^{2+}), cementation, and pozzolanic reactions (Ingles & Metcalf, 1972; Prusinski & Bhattacharja, 1999).

Ion exchange and flocculation are interdependent reactions that occur immediately upon mixing lime with the soil, resulting in immediate changes in soil properties. The free calcium in lime replaces adsorbed cations on clay minerals, reducing the size of the diffuse double layer of water surrounding clay particles. This closer proximity of clay particles results in flocculation or agglomeration (Ingles & Metcalf, 1972; Mitchell & Soga, 2005).

This process is almost instant and results in an increase in soil pH (above 12), due to the presence of calcium hydroxide in the mixture. Calcium ions prevent water from penetrating the voids of clay minerals, neutralize their negative charges, reduce interlayer spacing, and reorient clay minerals. This further promotes flocculation and cation exchange, reducing the soil's tendency to expand (Herzog & Mitchell, 1963).

The rise in soil pH causes the dissolution of clay silicates and alumina, resulting in a gel of hydrated calcium silicates around clay particles, which gradually crystallizes over time. After these phenomena, slower reactions occur, including cementation and pozzolanic reactions. These pozzolanic reactions yield cementing compounds (e.g., C-S-H and C-A-H), reducing soil expansion and contraction. This is due to the reaction of lime with water, silica, iron silicates, and amorphous aluminum silicates present in the soil. These cementing compounds contribute to the gradual increase in soil strength and structural stability over several years (Mitchell & Soga, 2005; Nalbantoğlu, 2004).

Silvani et al. (2020) conducted experiments involving compacted soil-lime blends with varying lime contents (2%, 4%, and 6%) and different dry unit weights. The findings indicated that soil swelling tends to decrease as both lime content and porosity increase, highlighting the influence of both factors on soil swelling.

Ferreira et al. (2018) examined the behavior of expansive soil when mixed with lime in proportions of 3%, 5%, and 7% by weight. The addition of lime into the soil led to the flocculation and a reduction in the plasticity index. Specifically, the addition of 5% hydrated lime to the soil achieved effective stabilization, significantly reducing both the free swelling potential and swelling pressure to virtually zero values.

Phanikumar et al. (2015) observed a reduction in the Free Swell Index (FSI) as lime was progressively added to the blends. Specifically, as the lime content increased from 0% to 4%, the FSI decreased from 185% to 63.63%, while the swell potential decreased from 20% to 11.5%.

Phanikumar (2009) analyzed an expansive soil with a measured swell potential of 27%. The findings indicated a reduction of 48% and 89% in swell potential when the lime content was 2% and 4%, respectively. Additionally, it was observed that the swelling pressure decreased from an initial value of 330 kPa to 160 kPa with the addition of 4% lime content.

In a study conducted by Al-Rawas et al. (2005), free swelling tests were performed on expansive soil samples, yielding a free swell percentage of 9.39%. Upon the inclusion of 3% lime content, the FSI reduced to 2.7%. Further increasing the lime content to 6% reduced the FSI to zero.

Cement

Portland cement is typically manufactured through the calcination of limestone and clay at temperatures as high as 1400°C. Despite it has been widely used in swelling soil stabilization, the high costs and environmental impacts stemming from greenhouse gas emissions during its production, there is a growing need to replace it with more sustainable and eco-friendly alternatives (Araújo et al. 2023).

The mechanism by which Portland cement stabilizes swelling soils closely resembles the lime stabilization of these soil types. This process involves cation exchange, flocculation, hydration, and pozzolanic reactions (Ingles & Metcalf, 1972).

When Portland cement is introduced to wet swelling soil, it elevates the pH to 12.4 and releases calcium cations (Ca^{2+}). These cations replace some of the monovalent cations present in the soil, such as sodium (Na^+) and potassium (K^+), leading to soil flocculation. This phenomenon reduces the soil's susceptibility to water, thereby reducing its expansion (Mitchell & Soga, 2005).

Results found by Consoli et al. (2021) reveal the flocculation effect observed when Portland cement is added to swelling soil. In a blend comprising 50% bentonite and 50% kaolin, the swelling exceeds 55%. However, with the addition of 4% Portland cement to the blend, the swelling rapidly decreases to 31.5%.

Simultaneously, cementitious hydration occurs, resulting in the formation of calcium aluminum hydrate. These hydration products initially have a gelatinous consistency, but with curing time, they become stiffer and significantly contribute to the improvement of soil strength. As depicted in Figure 1, elaborated by authors' data, the increase in Portland cement content leads to an increase in the unconfined compressive strength (UCS) of a smectite-rich soil. However, this improvement is relatively less pronounced compared to the increase observed in UCS results when lime is added. This difference in UCS is likely due to the more effective soil flocculation promoted by lime.

The hydration of Portland cement generates lime as a byproduct. This lime can engage in a pozzolanic reaction with silica and alumina if they are present in the blend, thereby enhancing soil resistance, as previously discussed (Ingles & Metcalf, 1972).

Phanikumar et al. (2015) demonstrated that with the addition of cement into the blends, the Free Swell Index (FSI) consistently decreased. In this context, the FSI decreased from 185% to 110%, while the swell potential also exhibited a decrease, dropping from 20% to 10.5% as the cement content was increased from 0% to 20%.

Likewise, Al-Rawas et al. (2005) reported that at a 3% cement content, the FSI reduced to 5.43% compared to untreated soil. When the cement content was increased to 6%, the FSI decreased further to 4.57%, and with 9% cement content, the FSI decreased even more to 3.67%.

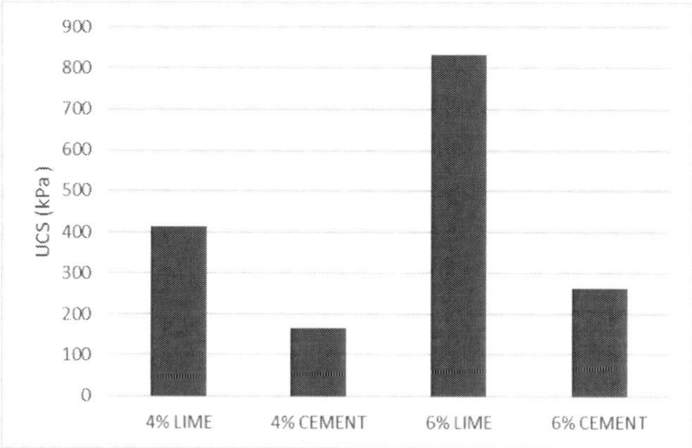

Figure 1. Unconfined compressive strength of a swelling smectite soil stabilized with lime and Portland cement with 28 days of curing.

Alternative Solution for Swelling Soil

Lime and cement have traditionally proven effective in mitigating swelling soil problems. However, these materials raise environmental concerns due to their manufacturing processes. These concerns involve high carbon dioxide emissions, increased energy consumption, and the exhaustion of non-renewable natural resources, which have been intensified by their overconsumption in recent decades (Ahmed et al. 2021; Blayi et al. 2021; Firoozi et al. 2017). Another conventional approach to improve swelling soil involves partial replacement with sand. However, this method also presents environmental implications, as sand is a non-renewable material (Ma et al. 2018; Phanikumar et al. 2021).

Among these environmental impacts, carbon dioxide (CO2) emissions from cement production are a contributor to global warming, accounting for nearly 8% of global CO2 emissions. For instance, in 2019, global cement production exceeded 4.1 billion tons (Guo et al. 2023). To diminish environmental impacts effectively, it is crucial to seek innovative, sustainable, and green solutions to replace Portland cement/lime and promote the adoption of these alternative materials.

Research conducted globally has explored the use of industrial residues as partial replacements for expansive soils, presenting an environmentally friendly approach to address soil swelling. This approach not only helps control soil swelling but also contributes to mitigating the issue of waste disposal (Firat et al. 2017; Ikeagwuani et al. 2019; Yilmaz & Yurdakul, 2017).

Several studies have investigated the effectiveness of utilizing industrial and organic by-products as solutions to these issues. Prior research has shown that non-traditional materials such as sugarcane bagasse ash (SCBA), rice husk ash (RHA), fly ash (FA), and electric arc furnace slag (EAF) are efficient additives with pozzolanic reactivity (Cordeiro et al. 2009; Horpibulsuk et al. 2011; Jamil et al. 2013; Silvani et al. 2023). Additionally, rock dust waste (RDW) and polyethylene terephthalate (PET), while non-reactive materials, serve as effective additives and act as significant stabilizers for granulometric purposes. These materials show potential as efficient stabilizing agents, whether used individually or, even more effectively, in combination with other additives (Barman & Dash, 2022; Seco et al. 2011).

Regarding their impact on swelling behavior, these materials can provide volumetric stability in the blends. However, in terms of achieving strength gains through pozzolanic reactions, these materials alone cannot yield a fully cemented structure. To achieve this, they need to be combined with a sodium or calcium-rich material, such as conventional cementing agents (although in lower proportions compared to when used on their own) or alternative options like carbide lime and eggshell lime (Consoli et al. 2020; Horpibulsuk et al. 2011).

Furthermore, brackish water serves as a catalyst for pozzolanic reactions, due to its high salt content, which leads to a rapid increase in strength during the early curing days. Additionally, it can serve as a feasible alternative to potable water, which are scarce assets in the world (Silvani et al. 2023). Therefore, a diverse range of materials can be used as total or partial replacements for high greenhouse gas emissions materials.

Reactive Materials

As previously discussed, the use of pozzolanic materials can effectively reduce swell variation while also enhancing mechanical properties over time. This mitigation occurs due to changes in the chemical and mineralogical properties of the soil-mixtures. The following section presents results obtained from swelling soils treated with various unconventional materials.

Rice Husk Ash – RHA

Rice husk is a byproduct from rice crop cultivation, typically yielding one ton of rice husk from the production of four tons of rice. This waste product is frequently incinerated, generating rice husk ash (RHA) equivalent to nearly 20% of its weight. The disposal of this ash is often inadequate, raising environmental concerns. Its lightweight nature makes it easily carried by the wind, leading to groundwater table contamination. Nevertheless, the RHA is a valuable pozzolanic material due to its high content of amorphous silica, exceeding 80% (Siddika et al. 2021).

Jain et al. (2020) investigated how RHA influences the swelling behavior and strength characteristics of Black Cotton Soil (BCS), an expansive soil found in India. They varied levels of RHA content from 0% to 20%, and their findings provided valuable insights into the physical properties affected by RHA. The authors attributed the mitigation of free swell index with increasing RHA content to the replacement of soil particles by non-plastic and non-swelling particles present in RHA. Additionally, the replacement of monovalent cations in BCS with multivalent cations like Ca^{2+} and Fe^{2+} found in RHA led to a reduction in the size of the double diffuse layer and a decreased tendency for expansion.

Additionally, Onyelowe et al. (2021) conducted an expansion analysis using RHA in a swelling soil, adding 1% of the soil's dry mass of various additives (CaO, $Ca(OH)_2$, and $CaCO_3$). The study revealed that as the RHA content increased, the free swell index decreased, regardless of the type of additive used. This reduction in soil swelling was substantial, with a reduction of more than four times.

Regarding strength characteristics, Jain et al. (2020) carried out unconsolidated undrained (UU) triaxial tests on BCS stabilized with RHA. Their findings indicated that the peak deviatoric stress increased with higher RHA content, despite of the applied confining pressure. This increase is likely due to granulometric stabilization and a potentially higher number of interlocked particles. However, no distinct pattern in cohesion related to the cementitious matrix was observed.

Ma et al. (2020) carried out an analysis of UCS over time in expansive soil with an 80/20 proportion of lime/RHA. Their results showed a greater improvement in strength in mixtures with higher RHA and lime content. This behavior can be attributed to the pozzolanic reactions responsible for the formation of C-S-H, which contributes to the blend's strength. However, after reaching a 15% lime/RHA content, no considerable additional strength improvement was observed, suggesting that higher contents may not be as effective.

Sugarcane Bagasse Ash – SCBA

Sugarcane bagasse ash (SCBA) shares similar properties with RHA, as it is also an industrial byproduct. In the sugar cane production, the bagasse is a residue generated during this process, with approximately 95% of this bagasse being incinerated in the boilers of sugar mills to generate energy. For instance, during the Brazillian 2020/2021 crop, approximately 657.4 million tons of bagasse were generated to produce sugar and ethanol. This bagasse undergoes calcination through uncontrolled burning temperatures, resulting in the formation of SCBA as a byproduct of this combustion process. Annually, up to 10 million tons of SCBA are generated (Cordeiro et al. 2009; Franco-Luján et al. 2021).

The improper disposal of SCBA in the environment is an ongoing concern that requires careful and appropriate management, which is not always implemented. This can lead to environmental problems and risks, primarily due to the presence of potassium and other heavy metals in its composition, which can pollute the soil and groundwater. Additionally, the fine fraction of SCBA can be easily dispersed in the air, causing respiratory risks to people residing near areas where the ash is improperly disposed (Brinkman et al. 2018; Cordeiro et al. 2019; Le Blond et al. 2014).

SCBA contains high amorphous silica and alumina content, as well as calcium and magnesium oxides, which are essential elements that trigger cementitious and pozzolanic reactions (see Chapter 5). This turns the SCBA into a viable green stabilizer for ground improvement through chemical techniques (Cordeiro et al. 2019). Recent research has highlighted SCBA's potential as a chemical stabilizer for expansive soils, yielding its effectiveness in controlling soil expansion and improving mechanical properties (Al-Kalili et al. 2022; Khan et al. 2021; Wubshet & Tadesse, 2014).

Silvani et al. (2023) conducted a comprehensive study on the volumetric behavior of a swelling soil stabilized with varying SCBA contents, ranging from 0% to 25%. By adding 6.25% of SCBA, little to no swelling was observed, and with further additions of 12.50% or more, the free swelling reduced to less than 0.5%, achieving volumetric stability. Furthermore, the authors compared the normalized vertical swelling over time to studies that used Portland cement and conventional hydrated lime. The results indicated that 6.25% of SCBA exhibited similar behavior to Portland cement and hydrated lime, with higher SCBA contents outperforming these conventional cementing agents in terms of reducing swelling.

Regarding the chemical and mineralogical changes after adding SCBA into the swelling soil, the same authors conducted Methylene blue index tests to investigated cation exchange capacity (CEC). The results revealed that adding 6.25% of SCBA or more to the swelling soil reduced CEC by more than two times. However, adding more than 12.5% did not result in further improvement. The authors stated that the addition of SCBA caused a major reduction in specific surface area and water absorption potential, indicating the formation of new phases composed of coarser particles. This analysis by Silvani et al. (2023) was based on the cations exchange from SCBA (high magnesium - Mg^{2+}) and calcium (Ca^{2+}) to monovalent cations present in thc swelling soil (Na^{+}, K^{+}).

In terms of mechanical strength behavior, Silvani et al. (2023) performed a series of unconfined compressive strength (UCS) tests on a swelling soil stabilized with SCBA, cement, tap water, and brackish water. The addition of brackish water improved strength gain during the early curing period compared to tap water. This improvement is due to the high salt content in brackish water, which acts as a catalyst for pozzolanic reactions. This finding suggests that brackish water could serve as a sustainable alternative to tap water or potable water, which is a valuable and scarce resource, in the stabilization of swelling soil-SCBA blends. Furthermore, the addition of both SCBA and cement to the blends led to an increase in soil strength, regardless of the type of water used in the mixture.

Wubshet & Tadesse (2014) conducted tests to assess the impact of different additives on an expansive soil, individually and in combination. Their findings revealed that uncured samples of the expansive soil exhibited reduced free swelling values. The initial value of 11.83% decreased to 6.07% with the addition of 3% lime, to 7.56% with 15% bagasse ash, and to 2.15% with a combination of 3% lime and 15% bagasse ash. Following a 7-day curing period, the samples exhibited further reductions, decreasing from the natural soil value of 11.83% to 0.95% with 3% lime, to 6.78% with 15% bagasse ash, and to 0.26% with a combination of 3% lime plus 15% bagasse ash.

Likewise, Dang et al. (2021) conducted tests in which soil specimens were prepared by varying the proportions of SCBA and SCBA+Lime over a wide range from 0% to 25% and lime content ranging from 0% to 6.25% by the dry weight of soil. The results revealed a significant improvement in various soil properties upon the addition of combined SCBA-lime. Specifically, there was an impressive 815% increase in maximum compressive strength, a notable 9.2-fold improvement in bearing capacity, a complete elimination of free swell potential, and a substantial 83% reduction in compressibility properties in the stabilized soils.

Fly Ash – FA

In alignment with the trends discussed earlier, Nalbantoğlu (2004) examined the effectiveness of Class C fly ash (FA) as a stabilizer for expansive soils. Fly ash is primarily produced from coal combustion in thermal power stations for electricity generation. Class C FA is rich in calcium and lime, exceeding 16% content. This attribute gives FA self-cementing properties, making it an economical and high-quality stabilizing agent.

Nalbantoğlu (2004) used Class C FA in two different expansive soils. The author found that adding 15% of FA into the soil significantly reduced swelling. Further increases in the FA content to the soil did not yield improvements and, in fact, worsened the performance. The curing time, however, played a crucial role assessing FA's behavior. Specimens subjected to nearly 100 days of curing exhibited the best performance in mitigating swelling, regardless of the specific soil type. The author attributed this behavior to the cementing characteristics of FA and the potential influence of pozzolanic reactions in reducing swelling.

Regarding chemical and mineralogical composition after the addition of FA, the same author conducted CEC tests, revealing a similar trend to that observed by Silvani et al. (2023). When FA was added, the CEC tended to decrease, indicating a change in mineralogy that no longer resembled the untreated soil. New mineralogical phases emerged, playing a crucial role in mitigating swelling (Nalbantoğlu, 2004).

Additionally, Zha et al. (2008) conducted a study on swelling soil, examinig the impact of FA-lime on the FSI. The authors concluded that the addition of FA to the blends can reduce the lime content required to achieve the desired volumetric stability. Hence, when combined with low lime contents, a high FA content (approximately 15%) can be as effective as higher lime contents. This trend offers valuable insights for reducing environmental impacts associated with lime application in the field.

Notably, to the author's knowledge, there is a limited body of literature addressing the mechanical properties of FA-swelling soil blends. Phanikumar (2009) conducted a comparative study that evaluated the use of lime and FA for stabilizing expansive clay through UCS tests. However, the results were notably affected by the suction effect (specimens were not submerged in water for saturation), preventing a definitive conclusion. Therefore, it is evident that further research and data are required in the literature to better understanding of the mechanical properties of swelling soil-FA blends, particularly on aspects such as strength and stiffness.

Electric Arc Furnace – EAF

Another non-conventional reactive material that can effectively address swelling soil issues is the electric arc furnace slag (EAF). EAF is an industrial byproduct generated during the steel production process. Similar to the alternative materials mentioned earlier, EAF is a no-cost material as it originates from industrial waste streams. Moreover, it helps alleviate concerns associated with landfilling, which can induce several environmental issues. Therefore, the utilization of EAF, along with the other before-mentioned materials, can reduce environmental impacts and offer a sustainable solution for geotechnical applications (Wu et al. 2019).

Shahsavani et al. (2021) conducted a comprehensive study focusing on highly expansive bentonite stabilized with EAF and nano-silica, with consideration for cold regions. The results presented regarding swelling potential indicate a reduction as the EAF content increases, up to 20%. The authors also emphasized the significant influence of curing time on swelling potential, with a notable decrease observed in longer curing periods. This behavior aligns with findings from Nalbantoglu (2004), that it was used Class C FA. In addition, the study revealed a 60% increase in unconfined compressive strength with higher EAF content, particularly in extended curing periods. Therefore, regardless of the alternative material content used, curing time stands out as a critical parameter when assessing swelling and strength behaviors of stabilized expansive soils.

The same authors conducted a durability analysis using the freeze-thaw cycles methodology to simulate conditions in cold regions. It is important to highlight that expansion is denoted by a positive sign (+), while contraction/shrinkage is represented by a negative sign (-). During the freezing process, volumetric expansion occurs due to the formation of ice crystals, while a reduction in volume corresponds to the ice's melting (Kamei et al. 2012). The maximum volumetric strain obtained for the EAF-stabilized soil was 2.75%, whereas non-stabilized swelling soil exhibited the highest volumetric expansion at 6%. The authors emphasized the positive impact of EAF on the durability of expansive soils, particularly in the initial cycles.

In their extensive experimental program, Shahsavani et al. (2021) also conducted a micro-fabric analysis after introducing EAF to swelling soil-nano-silica blends. The magnifications revealed distinct fabric patterns. The EAF addition led to the transformation from a weak and dispersed fabric to a flocculated structure, complemented by the formation of a cementitious phase. The voids within the bentonite reduced considerably with the introduction of EAF, attributed to the formation of gel-like materials that bonded the particles together.

To conclude their proposed paper, the authors presented XRD (X-ray diffraction) data for both bentonite and treated bentonite. The peaks associated with montmorillonite and quartz in bentonite experienced a considerable reduction upon the introduction of these alternative materials. Additionally, new peaks of C-S-H phases emerged, indicative of the cementitious phase. These collective findings provide a comprehensive understanding of how EAF and other pozzolanic materials with similar characteristics perform under swelling soil conditions.

Non-Reactive Materials

Rock Dust Waste (RDW)

Rock Dust Waste (RDW) is a fine-grained rock residue generated from the manufacturing processes of various ornamental rocks. It resembles sand in its texture and is often improperly disposed of in tailings dams (Balkis, 2017; Lucena et al. 2023; Silva et al. 2005). RDW can serve as a partial replacement additive in swelling soils treatment, providing a sustainable and cost-effective solution. Its use reduces the demand for sand exploration from natural sites and mitigates environmental impacts (Chaitanya & Krishna, 2018; Fasihnikoutalab et al. 2017; Latifi et al. 2017).

Studies have utilized different types of rock dust for swelling soil stabilization. For instance, Bapiraju & Prasad (2019) incorporated RDW at contents ranging from 5% to 30% into an expansive soil and observed a reduction in swelling potential, ranging from 7.0% to 1.17%, along with improvements in soil properties. Abdelkader et al. (2022) employed granite dust (also RDW) at contents of 5% to 30%, resulting in a decrease in swelling potential from 6.63% to 1.54%. Ogila (2016) demonstrated that the addition of ornamental limestone dust (other type of RDW) to highly expansive soil significantly reduced the percent of free swell up to 82.9%.

Lucena et al. (2023) suggested that RDW is likely to remain chemically inert due to its crystalline nature and may not react with the expansive soil. From their findings in unconfined compressive strength (UCS) and X-ray diffraction tests, they proposed that RDW may not exhibit pozzolanic reaction with the soil.

However, the authors noted that the RDW content did influence the swelling behavior of soil-RDW mixtures, with higher RDW contents resulted in reduced vertical swelling for the same porosity. While the expansive soil initially had free swelling ranging from 4.61% to 8.14% (across three different dry unit weights), the addition of 50% RDW into the soil reduced the free swelling to a range of 3.08% to 5.4%, depending on the dry unit weight value (Lucena et al. 2023).

Polyethylene Terephthalate (PET)

Nowadays, the issue of plastics and microplastics has become increasingly pressing. According to Sharma et al. (2020), plastic production is projected to reach a staggering 33 billion tons by 2050. Additionally, Wang et al. (2022) estimated that approximately 80% of the world's plastic waste ends up into landfills.

Polyethylene terephthalate (PET) is a commonly used type of plastic, particularly in the production of liquid containers like bottles. Micronized PET refers to a finely powdered variant of PET, obtained by grinding used PET plastic bottles into particles typically ranging from 0.2 mm to 0.075 mm (Ghabchi et al. 2021).

Martínez et al. (2016) and Louzada et al. (2019) found that micronized PET can improve the friction angle of clayey soil. However, the confining stress is a crucial parameter for improving the interaction between the soil and PET. The authors explained that grain-size improvement occurs at higher confining stresses (i.e., fewer voids and thus a greater demand for crushed PET), while at lower confining stresses, PET is unable to fill the voids in the soil. The best results were obtained with mixtures containing 30% and 5% of PET, whereas 3% of PET did not perform as effectively as the other proportions.

Carvalho et al. (2019) investigated the resilient modulus for blends of clayey soil and micronized PET to assess its suitability for pavement bases.

As a result of their positive findings, there is a reduction in the base layer thickness in pavement design for the same maximum project duration. However, the effectiveness strongly depends on the PET content, with higher PET contents (>5%) proving less effective than lower contents (as used in their research at 3%).

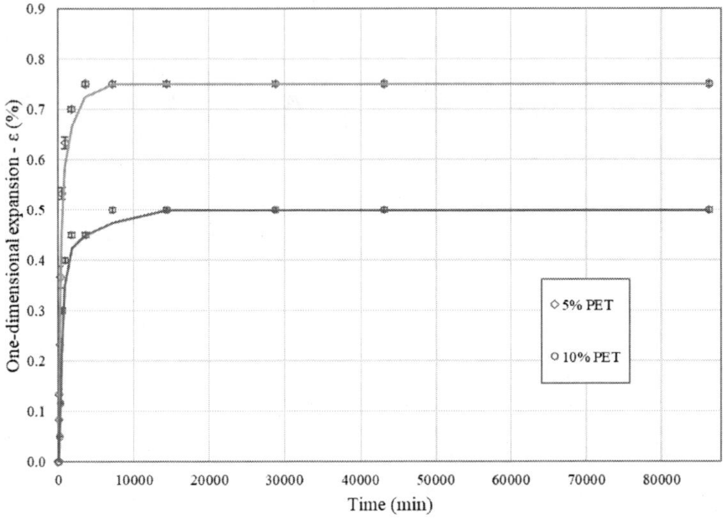

Figure 2. One-dimensional expansion over time for soil/lime/PET blend with 4% lime and dry unit weight of 15 kN/m³.

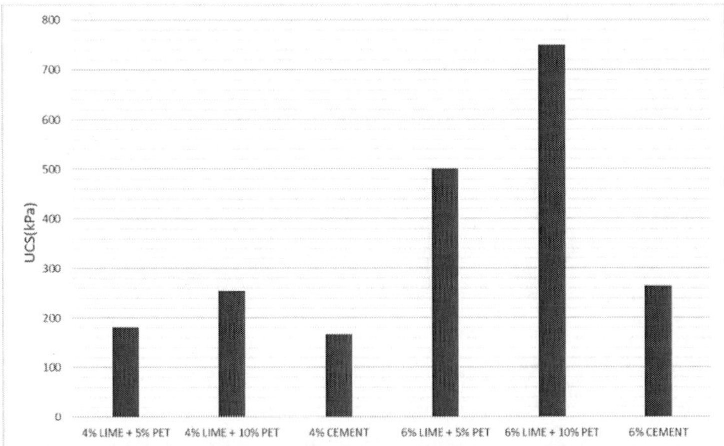

Figure 3. Unconfined compressive strength of expansive soil stabilized with different materials.

Figure 2 depicts the one-dimensional expansion of smectite soil stabilized with 4% lime carried out by the authors, a dry unit weight of 15 kN/m³, and micronized PET contents of 5% and 10%. An increase in PET content leads to a reduction in soil swelling. Similar results were obtained for different lime contents and dry unit weights. This occurrence can be attributed to the inert and non-expansive characteristics of micronized PET. As the proportion of micronized PET in the mixture increases, it replaces the expansive clay components, resulting in reduced swelling.

The UCS of swelling soil can also be enhanced by combining micronized PET and lime. Figure 3 illustrates the UCS of swelling soil (rich in smectite) stabilized with a lime/micronized PET mixture and Portland cement, performed by the authors. According to Figure 3, it is more effective to use 4% lime in combination with 5% micronized PET than to use 4% Portland cement. Similar results are observed when comparing 6% of Portland cement with a mixture of 6% lime and 5% micronized PET. Notably, an increase in micronized PET content leads to higher UCS while maintaining the same lime content.

Therefore, micronized PET can serve as a valuable additive in conjunction with lime for swelling soil stabilization, enabling the use of lower contents of traditional binders.

Final Considerations

Swelling soils represent a geotechnical challenge found in several regions worldwide. Their significant volumetric changes may lead to structural risks, especially for lightweight constructions.

Traditionally, stabilizing swelling soils with lime or Portland cement has been a common technique. This approach allows the use of local soil resources while minimizes the need for soil disposal areas. However, the production of these materials carries a heavy environmental burden, including high carbon dioxide emissions, increased energy consumption, and the depletion of non-renewable natural resources.

In this chapter, the researchers presented the feasibility of using industrial residues as stabilizers for expansive soils, offering an environmentally friendly approach to address soil swelling. Reactive materials such as sugarcane bagasse ash (SCBA), rice husk ash (RHA), fly ash (FA), and electric arc furnace slag (EAF) have proven to be effective additives, inducing soil flocculation and cementation through the pozzolanic reaction. On the other hand, non-reactive materials like rock dust waste (RDW) and polyethylene terephthalate (PET) can replace the expansive clay components, resulting in reduced swelling.

It is noteworthy that incorporating waste materials into soil stabilization can provide an alternative purpose for industrial waste products, diverting them from improper disposal and contributing to sustainable geotechnical practices.

References

Abdelkader, H. A. M., Ahmed, A. S. A., Hussein, M. M. A., Ye, H. & Zhang, J. (2022). An experimental study on geotechnical properties and micro-structure of expansive soil stabilized with waste granite dust. *Sustainability, 14(10)*, 6218. https://doi.org/10.3390/SU14106218.

Ahmed, M., Bashar, I., Alam, S. T., Wasi, A. I., Jerin, I., Khatun, S. & Rahman, M. (2021). An overview of Asian cement industry: Environmental impacts, research methodologies and mitigation measures. *Sustainable Production and Consumption, 28*, 1018–1039. https://doi.org/10.1016/J.SPC.2021.07.024.

Al-Kalili, A., Ali, A. S. & Al-Taie, A. J. (2022). A Review on expansive soils stabilized with different pozzolanic materials. *Journal of Engineering, 28(1)*, 1–18. https://doi.org/10.31026/j.eng.2022.01.01.

Al-Rawas, A. A., Hago, A. W. & Al-Sarmi, H. (2005). Effect of lime, cement and Sarooj (artificial pozzolan) on the swelling potential of an expansive soil from Oman. *Building and Environment, 40(5)*, 681–687. https://doi.org/10.1016/J.BUILDENV.2004.08.028.

Araújo, M. T. de, Ferrazzo, S. T., Chaves, H. M., Rocha, C. G. da & Consoli, N. C. (2023). Mechanical behavior, mineralogy, and microstructure of alkali-activated wastes-based binder for a clayey soil stabilization. *Construction and Building Materials, 362*, 129757. https://doi.org/10.1016/J.CONBUILDMAT.2022.129757.

Balkis, A. P. (2017). The effects of waste marble dust and polypropylene fiber contents on mechanical properties of gypsum stabilized earthen. *Construction and Building Materials, 134*, 556–562. https://doi.org/10.1016/J.CONBUILDMAT.2016.12.172.

Bapiraju, P. & Prasad, N. K. (2019). An experimental investigation on expansive soil in Conjunction with Egg Shell Powder and Rock Dust. International Journal of Advanced Research in Engineering and Technology (IJARET), 10(5), 9–21.

Barman, D. & Dash, S. K. (2022). Stabilization of expansive soils using chemical additives: A review. *Journal of Rock Mechanics and Geotechnical Engineering, 14(4)*, 1319–1342. https://doi.org/10.1016/J.JRMGE.2022.02.011.

Blayi, R. A., Sherwani, A. F. H., Mahmod, F. H. R. & Ibrahim, H. H. (2021). Influence of rock powder on the geotechnical behaviour of expansive soil. *International Journal of Geosynthetics and Ground Engineering, 7(1)*, 1–13. https://doi.org/10.1007/S40891-021-00260-3/METRICS.

Bolt, G. H. (2015). Physico-chemical analysis of the compressibility of pure clays. *Géotechnique, 6(2)*, 86–93. https://doi.org/10.1680/GEOT.1956.6.2.86.

Brinkman, M. L. J., Cunha, M. P. da, Heijnen, S., Wicke, B., Guilhoto, J. J. M., Walter, A., Faaij, A. P. C. & van der Hilst, F. (2018). Interregional assessment of socio-economic effects of sugarcane ethanol production in Brazil. *Renewable and Sustainable Energy Reviews, 88*, 347–362. https://doi.org/10.1016/j.rser.2018.02.014.

Carvalho, B., Casagrande, M. & Farias, M. (2019). Evaluation of resilient behavior of a clayey soil with polyethylene terephthalate (PET) insertion for application in pavements base. *E3S Web of Conferences, 92*, 12006. https://doi.org/10.1051/E3SCONF/20199212006.

Chaitanya, R. K. & Krishna, V. R. (2018). Experimental study on soil stabilization using waste fiber material. International Journal for Research in Applied Science & Engineering, 6(I). https://doi.org/10.22214/ijraset.2018.1172.

Consoli, N. C., Araújo, M. T., Ferrazo, S. T., Rodrigues, V. de L. & Rocha, C. G. da. (2021). Increasing density and cement content in stabilization of expansive soils: Conflicting or complementary procedures for reducing swelling? *Canadian Geotechnical Journal, 58(6)*, 866–878. https://doi.org/10.1139/CGJ-2019-0855.

Consoli, N. C., Caicedo, A. M. L., Saldanha, R. B., Scheuermann Filho, H. C. S. & Acosta, C. J. M. (2020). Eggshell produced limes: Innovative materials for soil stabilization. *Journal of Materials in Civil Engineering, 32(11)*, 6020018. https://doi.org/10.1061/(ASCE)MT.1943-5533.0003418.

Consoli, N. C., Marin, E. J. B., Samaniego, R. A. Q., Scheuermann Filho, H. C., Miranda, T. & Cristelo, N. (2019). Effect of mellowing and coal fly ash addition on behavior of sulfate-rich dispersive clay after lime stabilization. *Journal of Materials in Civil Engineering, 31(6)*, 4019071. https://doi.org/10.1061/(ASCE)MT.1943-5533.0002699.

Cordeiro, G. C., Andreão, P. V. & Tavares, L. M. (2019). Pozzolanic properties of ultrafine sugar cane bagasse ash produced by controlled burning. *Heliyon, 5(10)*, e02566. https://doi.org/10.1016/j.heliyon.2019.e02566.

Cordeiro, G. C., Toledo Filho, R. D. & Fairbairn, E. M. R. (2009). Effect of calcination temperature on the pozzolanic activity of sugar cane bagasse ash. *Construction and Building Materials, 23(10)*, 3301–3303. https://doi.org/10.1016/J.CONBUILDMAT.2009.02.013.

Dang, L. C., Khabbaz, H. & Ni, B. J. (2021). Improving engineering characteristics of expansive soils using industry waste as a sustainable application for reuse of bagasse ash. *Transportation Geotechnics, 31*, 100637. https://doi.org/10.1016/J.TRGEO.2021.100637.

Fasihnikoutalab, M. H., Asadi, A., Unluer, C., Huat, B. K., Ball, R. J. & Pourakbar, S. (2017). Utilization of alkali-activated olivine in soil stabilization and the effect of carbonation on unconfined compressive strength and microstructure. *Journal of Materials in Civil Engineering, 29(6)*, 06017002. https://doi.org/10.1061/(ASCE)MT.1943-5533.0001833.

Ferreira, S. R. de M., Paiva, S. C. de, Morais, J. J. O. & Viana, R. B. (2018). Avaliação da expansão de um solo do município de Paulista-PE melhorado com cal. *Matéria (Rio de Janeiro), 22*. https://doi.org/10.1590/S1517-707620170005.0266.

Firat, S., Khatib, J. M., Yilmaz, G. & Comert, A. T. (2017). Effect of curing time on selected properties of soil stabilized with fly ash, marble dust and waste sand for road sub-base materials. *Waste Management and Research, 35(7)*, 747–756. https://doi.org/10.1177/0734242X17705726.

Firoozi, A. A., Guney Olgun, C., Firoozi, A. A. & Baghini, M. S. (2017). Fundamentals of soil stabilization. *International Journal of Geo-Engineering, 8(1)*, 1–16. https://doi.org/10.1186/S40703-017-0064-9/FIGURES/3.

Franco-Luján, V. A., Mendoza-Rangel, J. M., Jiménez-Quero, V. G. & Montes-García, P. (2021). Chloride-binding capacity of ternary concretes containing fly ash and untreated sugarcane bagasse ash. *Cement and Concrete Composites, 120*, 104040. https://doi.org/10.1016/J.CEMCONCOMP.2021.104040.

Ghabchi, R., Dharmarathna, C. P., & Mihandoust, M. (2021). Feasibility of using micronized recycled Polyethylene Terephthalate (PET) as an asphalt binder additive: A laboratory study. Construction and Building Materials, 292, 123377. https://doi.org/10.1016/J.CONBUILDMAT.2021.123377.

Guo, Y., Luo, L., Liu, T., Hao, L., Li, Y., Liu, P., & Zhu, T. (2023). A review of low-carbon technologies and projects for the global cement industry. Journal of Environmental Sciences, 136, 682–697. https://doi.org/10.1016/J.JES.2023.01.021.

Herzog, A. & Mitchell, J. K. (1963). Reactions accompanying stabilization of clay with cement. *Highway Research Record*, 36. https://onlinepubs.trb.org/Onlinepubs/hrr/1963/36/36-008.pdf.

Horpibulsuk, S., Phetchuay, C. & Chinkulkijniwat, A. (2011). Soil stabilization by calcium carbide residue and fly ash. *Journal of Materials in Civil Engineering, 24(2)*, 184–193. https://doi.org/10.1061/(ASCE)MT.1943-5533.0000370.

Ikeagwuani, C. C. & Nwonu, D. C. (2019). Emerging trends in expansive soil stabilisation: A review. *Journal of Rock Mechanics and Geotechnical Engineering, 11(2)*, 423–440. https://doi.org/10.1016/j.jrmge.2018.08.013.

Ingles, O. G. & Metcalf, J. B. (1972). Soil stabilization principles and practice. Butterworth and Company Publishers Limited. https://trid.trb.org/view/137528.

Jain, A., Choudhary, A. K. & Jha, J. N. (2020). Influence of rice husk ash on the swelling and strength characteristics of expansive soil. *Geotechnical and Geological Engineering, 38(2)*, 2293–2302. https://doi.org/10.1007/S10706-019-01087-6/FIGURES/14.

Jamil, M., Kaish, A. B. M. A., Raman, S. N. & Zain, M. F. M. (2013). Pozzolanic contribution of rice husk ash in cementitious system. *Construction and Building Materials, 47*, 588–593. https://doi.org/10.1016/J.CONBUILDMAT.2013.05.088.

Kamei, T., Ahmed, A. & Shibi, T. (2012). Effect of freeze–thaw cycles on durability and strength of very soft clay soil stabilised with recycled Bassanite. *Cold Regions Science and Technology, 82*, 124–129. https://doi.org/10.1016/J.COLDREGIONS.2012.05.016.

Khan, F. K., Kantroo, S. & Sobti, J. (2021). A Short review on improvement in soil and cement properties by addition of bagasse ash. In *Ground Improvement Techniques* (pp. 73–83). Springer. https://doi.org/10.1007/978-981-15-9988-0_8.

Kulanthaivel, P., Soundara, B., Velmurugan, S. & Naveenraj, V. (2021). Experimental investigation on stabilization of clay soil using nano-materials and white cement.

Materials Today: Proceedings, 45, 507–511. https://doi.org/10.1016/j.matpr.2020.02.107.

Lambe, T. W. & Whitman, R. V. (1959). The role of effective stress in the behavior of expansive soils. Soil Mechanics Conference.

Latifi, N., Eisazadeh, A., Marto, A. & Meehan, C. L. (2017). Tropical residual soil stabilization: A powder form material for increasing soil strength. *Construction and Building Materials, 147*, 827–836. https://doi.org/10.1016/J.CONBUILDMAT.2017.04.115.

Le Blond, J. S., Tomatis, M., Horwell, C. J., Dunster, C., Murphy, F., Corazzari, I., Grendene, F., Turci, F., Gazzano, E., Ghigo, D., Williamson, B. J., Oppenheimer, C. & Fubini, B. (2014). The surface reactivity and implied toxicity of ash produced from sugarcane burning. *Environmental Toxicology, 29(5)*, 503–516. https://doi.org/10.1002/tox.21776.

Louzada, N. dos S. L., Malko, J. A. C. & Casagrande, M. D. T. (2019). Behavior of clayey soil reinforced with polyethylene terephthalate. *Journal of Materials in Civil Engineering, 31(10)*, 04019218. https://doi.org/10.1061/(ASCE)MT.1943-5533.0002863.

Lucena, L. C. de F. L., Silvani, C., Tenório, E. A. G. & Guimarães, S. G. (2023). Devising methodology to control clay swelling with rock dust waste and soil porosity. [Manuscript under review for *Ground Improvement*].

Ma, J., Su, Y., Liu, Y. & Tao, X. (2020). Strength and microfabric of expansive soil improved with rice husk ash and lime. *Advances in Civil Engineering, 2020*. https://doi.org/10.1155/2020/9646205.

Ma, Q. Y., Cao, Z. M. & Yuan, P. (2018). Experimental research on microstructure and physical-mechanical properties of expansive soil stabilized with fly ash, sand, and basalt fiber. *Advances in Materials Science and Engineering, 2018*. https://doi.org/10.1155/2018/9125127.

Martínez, I. M. R., Louzada, N. D. S. L., Repsold, L. M. & Casagrande, M. D. T. (2016). Mechanical behavior of reinforced clayey soil with fine crushed polyethylene terephthalate. *Key Engineering Materials, 668*, 404–410. https://doi.org/10.4028/WWW.SCIENTIFIC.NET/KEM.668.404.

Mitchell, J. K. & Soga, K. (2005). *Fundamentals of Soil Behavior* (3rd ed.). John Wiley & Sons.

Nalbantoğlu, Z. (2004). Effectiveness of class C fly ash as an expansive soil stabilizer. *Construction and Building Materials, 18(6)*, 377–381. https://doi.org/10.1016/J.CONBUILDMAT.2004.03.011.

Ogila, W. A. M. (2016). The impact of natural ornamental limestone dust on swelling characteristics of high expansive soils. *Environmental Earth Sciences, 75(24)*, 1–17. https://doi.org/10.1007/S12665-016-6305-Y/METRICS.

Olphen, H. Van. (1977). An introduction to clay colloid chemistry, for clay technologists, geologists, and soil scientists. In *An introduction to clay colloid chemistry, for clay technologists, geologists, and soil scientists*. (2nd ed., Issue 2nd edition). John Wiley and Sons Ltd.

Onyelowe, K. C., Onyia, M. E., Nguyen-Thi, D., Bui Van, D., Onukwugha, E., Baykara, H., Obianyo, I. I., Dao-Phuc, L. & Ugwu, H. U. (2021). Swelling potential of clayey soil modified with rice husk ash activated by calcination for pavement underlay by Plasticity Index Method (PIM). *Advances in Materials Science and Engineering, 2021*. https://doi.org/10.1155/2021/6688519.

Phanikumar, B. R. & Singla, R. (2016). Swell-consolidation characteristics of fibre-reinforced expansive soils. *Soils and Foundations, 56(1)*, 138–143. https://doi.org/10.1016/j.sandf.2016.01.011.

Phanikumar, B. R. (2009). Effect of lime and fly ash on swell, consolidation and shear strength characteristics of expansive clays: a comparative study. *Geomechanics and Geoengineering: An International Journal, 4(2)*, 175–181. https://doi.org/10.1080/17486020902856983.

Phanikumar, B. R., Dembla, S. & Yatindra, A. (2021). Swelling behaviour of an expansive clay blended with fine sand and fly ash. *Geotechnical and Geological Engineering, 39(1)*, 583–591. https://doi.org/10.1007/S10706-020-01480-6/METRICS.

Phanikumar, B. R., Sreedharan, R. & Aniruddh, C. (2015). Swell-compressibility characteristics of lime-blended and cement-blended expansive clays – A comparative study. *Geomechanics and Geoengineering, 10(2)*, 153–162. https://doi.org/10.1080/17486025.2014.902120.

Prusinski, J. R. & Bhattacharja, S. (1999). Effectiveness of Portland cement and lime in stabilizing clay soils. *Transportation Research Record, 1652(1)*, 215–227. https://doi.org/10.3141/1652-28.

Puppala, A. J., Manosuthikij, T. & Chittoori, B. C. S. (2013). Swell and shrinkage characterizations of unsaturated expansive clays from Texas. *Engineering Geology, 164*, 187–194. https://doi.org/10.1016/j.enggeo.2013.07.001.

Seco, A., Ramírez, F., Miqueleiz, L. & García, B. (2011). Stabilization of expansive soils for use in construction. *Applied Clay Science, 51(3)*, 348–352. https://doi.org/10.1016/j.clay.2010.12.027.

Shahsavani, S., Vakili, A. H. & Mokhberi, M. (2021). Effects of freeze-thaw cycles on the characteristics of the expansive soils treated by nanosilica and Electric Arc Furnace (EAF) slag. cold regions science and technology, 182, 103216. https://doi.org/10.1016/J.COLDREGIONS.2020.103216.

Sharma, M. D., Elanjickal, A. I., Mankar, J. S. & Krupadam, R. J. (2020). Assessment of cancer risk of microplastics enriched with polycyclic aromatic hydrocarbons. *Journal of Hazardous Materials, 398*, 122994. https://doi.org/10.1016/J.JHAZMAT.2020.122994.

Siddika, A., Mamun, M. A. Al, Alyousef, R., & Mohammadhosseini, H. (2021). State-of-the-art-review on rice husk ash: A supplementary cementitious material in concrete. Journal of King Saud University - Engineering Sciences, 33(5), 294–307. https://doi.org/10.1016/J.JKSUES.2020.10.006.

Silva, J. B., Hotza, D., Segadães, A. M. & Acchar, W. (2005). Incorporação de lama de mármore e granito em massas argilosas. *Cerâmica, 51(320)*, 325–330. https://doi.org/10.1590/S0366-69132005000400004.

Silvani, C., Guedes, J. P., Silva, J. C. da, Tenório, E. A. G. & Nascimento, R. (2023). Brackish water in swelling soil stabilization with lime and sugarcane bagasse ash (SCBA). *Soils and Rocks, 46(3)*, e2023010022. https://doi.org/10.28927/SR.2023.010022.

Silvani, C., Lucena, L. C. de F. L., Tenório, E. A. G., Scheuermann Filho, H. C. & Consoli, N. C. (2020). Key parameter for swelling control of compacted expansive fine-grained soil–lime blends. *Journal of Geotechnical and Geoenvironmental Engineering, 146(9)*, 06020012. https://doi.org/10.1061/(ASCE)GT.1943-5606.0002335.

Silvani, C., Silva, J. C. da & Guedes, J. P. C. (2023). Sugarcane bagasse ash as a green stabilizer for swelling soil. *Geotechnical and Geological Engineering, 1–12*. https://doi.org/10.1007/S10706-023-02628-W/FIGURES/7.

Snethen, D. R. (1979). Technical guidelines for expansive soils in highway subgrades. In *Report FHWA-RD-79-51* (p. 168). Federal Highway Administration, U.S. Department of Transportation.

Tiwari, N., Satyam, N. & Puppala, A. J. (2021). Strength and durability assessment of expansive soil stabilized with recycled ash and natural fibers. *Transportation Geotechnics, 29*, 100556. https://doi.org/10.1016/J.TRGEO.2021.100556.

Wang, C., Tang, J., Yu, H., Wang, Y., Li, H., Xu, S., Li, G. & Zhou, Q. (2022). Microplastic pollution in the soil environment: Characteristics, influencing factors, and risks. *Sustainability, 14(20)*, 13405. https://doi.org/10.3390/SU142013405.

Wu, J., Liu, Q., Deng, Y., Yu, X., Feng, Q. & Yan, C. (2019). Expansive soil modified by waste steel slag and its application in subbase layer of highways. *Soils and Foundations, 59(4)*, 955–965. https://doi.org/10.1016/J.SANDF.2019.03.009.

Wubshet, M. & Tadesse, S. (2014). Stabilization of expansive soil using bagasse ash & lime. *Zede Journal, 32*, 21–26. https://www.ajol.info/index.php/zj/article/view/142172.

Yilmaz, F. & Yurdakul, M. (2017). Evaluation of marble dust for soil stabilization. *Acta Physica Polonica A, 132(3)*, 710–711. https://doi.org/10.12693/APhysPolA.132.710

Zha, F., Liu, S., Du, Y. & Cui, K. (2008). Behavior of expansive soils stabilized with fly ash. *Natural Hazards, 47(3)*, 509–523. https://doi.org/10.1007/S11069-008-9236-4/FIGURES/14.

Chapter 4

Alkali-Activated Binders as an Alternative to Portland Cement: Chemistry, Properties, and Uses in Geotechnics

Rafael Francisco dos Santos[1,*]
Suéllen Tonatto Ferrazzo[2]
Marcelo Esposito[3]
and Eduardo Pavan Korf[1]
[1]Graduate Program in Environmental Science and Technology, Universidade Federal da Fronteira Sul, Erechim, RS, Brazil
[2]Graduate Program in Civil Engineering, Universidade Federal do Rio Grande do Sul, Porto Alegre, RS, Brazil
[3]Evironmental and Sanitary Engineering, Universidade Federal da Fronteira Sul, Erechim, RS, Brazil

Abstract

Portland cement is a traditional binder widely used in engineering applications such as concrete production and soil stabilization. As the world´s population has grown, demand for Portland cement has increased over the years, with more than 4.37 billion tons produced worldwide in 2021. However, this increase in consumption poses a major challenge because the production of Portland cement releases a large amount of CO_2 (g), representing more than 7% of global emissions of this gas, which is one of the main gases, responsible for global climate change. The production of conventional cement requires 1.5 tons of non-renewable raw materials and 3.5 GJ/ton of clinker. Recent studies suggest

---

* Corresponding Author's Email: rafael.francisco@uffs.edu.br.

In: Environmental Geotechnics
Editors: E. Pavan Korf, P. Domingos Marques Prietto, C. Müller et al.
ISBN: 979-8-89113-651-9

that global consumption of Portland cement could exceed 5 gigatons by 2050. This underscores the urgent need for a viable and sustainable alternative to this material, such as alkali-activated binders. These alternative cements are created through alkaline activation processes in which amorphous aluminosilicates (precursors) are dissolved in an alkaline solution. These alternative binders can be produced by one or two-part methods and in different systems, depending on the type of cementing components: high-calcium system, low-calcium system, and hybrid cement model. Alkali-activated binders are characterized by chemical resistance to acids and sulphates, resistance to high temperatures and the possibility of using a wide range of precursor materials compared to Portland cement. In recent years, several research have been carried out using industrial and agro-industrial wastes as a precursor and calcium source, such as fly ash, slags, glass waste, paper sludge, sugarcane bagasse ash, rice husk ash, red ceramic wastes, mining tailings, carbide lime, eggshell lime, etc. In the geotechnical engineering, these alkaline cements are used to stabilize natural soils, contaminated soils, mining tailings, reclaimed asphalt pavement, waste foundry sand, etc. The studies focus on the evaluation of mechanical behavior and microstructure. However, the leaching of contaminants and the sustainability of stabilization strategies are poorly understood.

Keywords: alkali-activated systems, characteristics and performance, waste valorization, stabilization/solidification

The Global Climate Change and its Relationship with the Cement Industry

Global climate changes have been observed since the middle of the 20th century and are mainly related to human activities, which are responsible for the emission of large quantities of gases associated with the greenhouse effect, especially CO_2, which is the main gas of this effect. It is evident that there are also natural processes (e.g., natural forest fires, volcanic activity, and among others) that release gases and thus further contribute to climate change. However, it is undeniable that humans are a major contributor to the emission of CO_2. Among the human activities responsible for releasing tons of CO_2, it is possible to highlight the industrial processes, especially the Portland cement industry, which account for more than 7% of global emissions of this gas (Belaïd, 2022).

Recent advances in the economies of emerging countries such as China, Russia, Brazil, India, and others have contributed to an increase the CO_2 emissions, making it extremely important to discuss alternatives to reduce greenhouse gas emissions (GHS). In 2020, at the height of the COVID-19 pandemic, CO_2 emissions decreased by about 5% compared to 2019 (EPA, 2023; Le Quéré et al., 2020). This decrease in CO_2 emissions during this period was directly related to government polices restricting human mobility, such as lockdown, ordered in various parts of the world. However, in the following years, after a temporary decrease, global CO_2 emissions increased again and were comparable to those before the pandemic, due to the reduction in containment and the increase in fossil fuel consumption (EPA, 2023; Le Quéré et al., 2020). On the other hand, despite the global decrease in CO_2 emissions during the pandemic, it is important to observe the opposite situation in terms of GHS emissions from the cement industry over the years, even during the pandemic period.

The global increase in cement production is inevitable and directly related to population growth and large demand in civil construction. A look at GHS emissions data shows that global CO_2 emissions from the cement industry were around 576 million tons in 1990. By 2014, this figure had increased to 2.083 billion tons (Benhelal et al., 2021). In fact, CO_2 emissions from the cement industry have continued to rise since 2014, alerting both researchers and governments to the urgent need to reduce these emissions and thereby mitigate the greenhouse effect. The International Energy Agency (IEA) has monitored CO_2 emissions per ton of cement production (Figure 1) and found that these emissions have increased over the year (Morgado et al., 2023). This represents a major challenge for the industry to overcome this situation and reduce CO_2 emissions in order to slow down global warming (Benhelal et al., 2021; Le Quéré et al., 2020).

Portland cement is essentially a mixture of limestone and clay (or shale) mixed with some additives and heated to 1450 °C to produce the clinker. During this process, the compounds of the raw materials form tricalcium silicate, dicalcium silicate, tricalcium aluminate and tetracalcium iron aluminate, which react with water and form bonds. When the raw material is heated, limestone (the majority of which consists of $CaCO_3$) decomposes endothermically, releasing large amounts of CO_2 into the atmosphere, according to the following chemical reaction 1.

$$CaCO_3 \rightarrow CaO + CO_2 \; \Delta H = +177.5 \text{ kJ mol}^{-1} \quad (1)$$

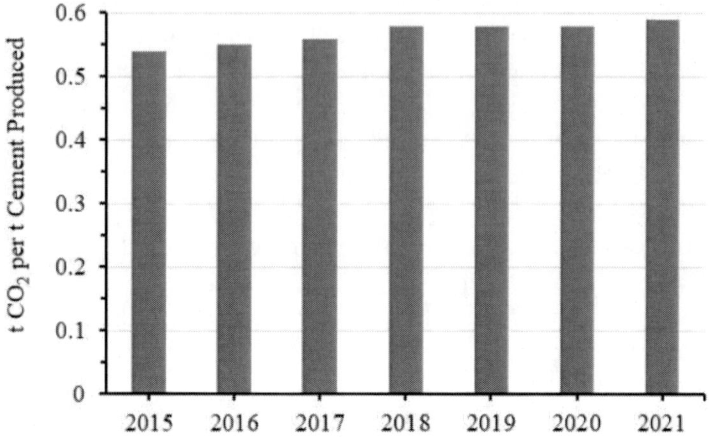

Figure 1. CO_2 emissions per ton of cement emitted by the global industry since 2015 to 2021, according to the International Energy Agency.

Unfortunately, the decomposition reaction of $CaCO_3$ in the production of Portland cement is necessary to obtain lime (CaO) in the mixture, and for each ton of limestone used to produce the cement, about 0.44 tons of CO_2 from $CaCO_3$ is released into the atmosphere. However, this amount of CO_2 does not correspond to the total emissions, as can be seen in Figure 1. It is necessary to consider other sources of GHS in the production chain, such as fossil fuels used in some industries to heat the furnaces for clinker production (Fernandez Pales & Leung, 2018; Morgado et al., 2023). According to the IEA, the production of one ton of cement in 2021 emitted about 0.59 tons of CO_2, with almost 50-60% of this gas coming from the thermal decomposition of limestone, while the other 30-40% corresponds to other CO_2 sources in the production chain (Benhelal et al., 2021). Recent estimates suggest that about 2.5 Gt of CO_2 will be released into the atmosphere by the cement industry by 2050 (Fernandez Pales & Leung, 2018; Morgado et al., 2023).

Despite the challenges facing the cement industry and the need to reduce greenhouse gas emissions, the production of Portland cement faces another major obstacle: competition with other industries for natural resources. The limestone used by the industry as a raw material, is mostly $CaCO_3$ (usually as calcite or aragonite) and is used not only in the cement industry, but also in agriculture to mitigate soil acidity, glass manufacturing, steel production, plastics manufacturing, water and wastewater treatment, and other applications. This industrial competition for a non-renewable natural resource with high pollution potential, when used for the cement industry has drawn attention to the development of sustainable alternative technologies to conventional cement to reduce dependence on the use of Portland cement, such as alkali-activated binders or alkaline cements (Monteiro et al., 2017).

Alkali-Activated Binders as Alternative to Portland Cement

The challenge of reducing CO_2 emissions from the cement industry to contribute to slowing global warming has highlighted the necessity to decrease reliance on Portland cement. This calls for the exploration of sustainable alternatives, such as alkali-activated binders (Benhelal et al., 2021; Provis, 2018; Shah et al., 2022). These alternative cements are produced by alkali-activation of a precursor material. A variety of materials can serve as precursors for alkali-activated binders, including industrial and agricultural wastes. This can result in a lower environmental impact compared to the conventional binders such as Portland cement (Garcia-Lodeiro et al., 2015a; Miller & Myers, 2020; Provis, 2018; Yusslee & Beskhyroun, 2022).

The alternative cements are prepared by the alkali-activation technique, which involves the reaction of an amorphous aluminosilicate as a precursor under alkaline conditions triggered by an alkali metal source (activator), resulting in a binder based on hydrous alkali-aluminosilicate and/or alkali-alkali earth aluminosilicate phases (Provis, 2018). These alternative binders are prepared by two main pathways: "one-part" (mixture of precursor and alkali activator in solid state) and "two-part" (mixture of precursor in solid state and alkali activator in liquid state). The "two-part" method was the predominant method in the early stages of alkali activation implementation in various fields, and most existing products are typically manufactured using this method. (Provis, 2018).

Depending on the type of cementitious constituents (CaO-SiO_2-Al_2O_3 system), alkali-activated binders are divided into two categories: (*i*) high-calcium cements and (*ii*) low-calcium cements, also called geopolymers in some cases. Each category has a different activation pattern, with moderate alkaline conditions required for high-calcium cements and strong alkaline conditions required for low-calcium cements. It is important to highlight that the two alkaline activated binders mentioned above have different chemical composition and structure (Choeycharoen et al., 2022; Garcia-Lodeiro et al., 2015).

For the alkali-activated binder with high-calcium content, prepared with materials such as blast furnace slag ($SiO_2 + CaO > 70\%$) and activated under moderate alkaline conditions, the main product of this reaction is calcium aluminosilicate hydrate gel (C-A-S-H), a similar product formed during the hydration of Portland cement. The C-A-S-H is produced by reacting Ca_2SiO_5 (typically represented as C_2S) and Ca_3SiO_5 (typically represented as C_3S) with water (Garcia-Lodeiro et al., 2015; Harrisson, 2019). Both chemical reactions (stoichiometric coefficients were intentionally omitted) are shown in Equations 2 and 3:

$$Ca_2SiO_3 + H_2O \rightarrow CaO\text{-}SiO_2\text{-}H_2O \quad (2)$$

$$Ca_3SiO_5 + H_2O \rightarrow CaO\text{-}SiO_2\text{-}H_2O + Ca(OH)_2 \quad (3)$$

Equations 2 and 3 are usually expressed as reaction 4 and 5 (unbalanced reaction), respectively:

$$C_2S + H \rightarrow C\text{-}A\text{-}\ S\text{-}H \quad (4)$$

$$C_3S + H \rightarrow C\text{-}A\text{-}\ S\text{-}H + CH \quad (5)$$

For alkali-activated binders prepared with low-calcium materials such as fly ash or metakaolin, an aggressive alkaline condition is required to favor the reaction to form the main product, such as sodium or potassium aluminosilicate hydrate gel (N-A-S-H or K-A-S-H) (Garcia-Lodeiro et al., 2011). The sodium aluminosilicate hydrate (N-A-S-H) gel is formed when NaOH is used as an alkaline activator. However, when KOH is used as the hydroxide in the reaction, potassium aluminosilicate hydrate (K-A-S-H) is formed. Apart from the possibility of using NaOH or KOH to prepare the geopolymer, the first one is much more commonly used in this reaction to prepare the alkali-activated binder (Li et al., 2020).

The N-A-S-H or K-A-S-H are inorganic polymers (geopolymer) and are generally described as a three-dimensional aluminosilicate network, which can be considered as a zeolite precursor and consists of disordered aluminosilicate tetrahedra (Garcia-Lodeiro et al., 2015; Li et al., 2020). These inorganic polymers are composed of AlO_4 and SiO_4 tetrahedra connected by bridging oxygen, with alkali ions, Na^+ or K^+, weakly bound to the structure (ionic bonding) and cancelling the charge of AlO_4^- (Wan et al., 2020). The general chemical reaction mechanism of N(K)-A-S-H formation proposed by V. Glukhovsky is divided into three stages: (*i*) destruction-coagulation, (*ii*) coagulation-condensation and (*iii*) condensation-crystallization (García-Lodeiro et al., 2012; Marvila et al., 2021).

In the first stage (destruction-coagulation), the HO^- ion from the alkaline mixture reacts with Si-O-Si and the bonds are broken, leading to the formation of SiOH and SiO^- Na^+. As can be observed, the alkaline cation (Na^+ or K^+) neutralizes the charge of the anion SiO^-. While the HO^- react with Si-O-Si, the reaction with the Al-O-Si bond occurs simultaneously in the same way. The HO^- attack the Al of the Al-O-Si and subsequently an internal attack of HO^- on the Si occurs, while the O-Si bond is broken, leading to the formation of complexes of $Al(OH)_4^-$. In the second stage (coagulation-condensation), coagulation of the disaggregated products occurs, leading to a polycondensation structure. The hydroxyl (HO^-) acts as catalyst and the clusters formed by the polymerization reaction produces colloidal particles. In the final stage (condensation-crystallization), the particles in the solid phase support the precipitation of the products and form the geopolymer (García-Lodeiro et al., 2012).

In addition to the C-A-S-H gel (from cements with high-calcium content) and the N(K)-A-S-H gel (from cements with low-calcium content), there is another model that produces a different binder known as hybrid alkali-activated cements (HAACs) or simply blended cements. The HAACs consist of a mixture of Portland cement (with a low content) with a high content of additional cementitious materials and an alkali activator. These products are formed by the alkali activation of materials with CaO, SiO_2, and Al_2O_3 content of about 20% (Garcia-Lodeiro et al., 2015; Mohapatra & Pradhan, 2022). Examples of hybrid cement systems include mixtures of Portland cement with industrial wastes such as blast furnace slag, phosphorous slags, fly ash, steel mill slag, and blast furnace slag or fly ash-blast furnace slag blends (García-Lodeiro et al., 2012).

The HAACs have attracted attention as a potential alternative binder due to their physical properties achieved by the coexistence of C-A-S-H gel and geopolymer gel from N(K)-A-S-H. In general, the literature has indicated that the workability and setting time of HAACs are generally lower than conventional Portland cement. In addition, mortars and concretes made with this alternative binder have historically achieved higher early strength compared to conventional Portland cement (Mohapatra & Pradhan, 2022; Park et al., 2022). Furthermore, High-Alumina Alkali-Activated Cements (HAACs) release 44% to 70% less CO_2 into the atmosphere than concrete produced with Portland cement (Amer et al., 2021; McLellan et al., 2011; Mohapatra & Pradhan, 2022). HAACs are complex blends initially consisting of CaO, SiO_2, and Al_2O_3 content greater than 20% and composed into different gels depending on the reaction condition. The N-A-S-H and C-S-H gels are not formed separately, but they interact with each other, changing their structure and composition (Garcia-Lodeiro et al., 2015).

The Main Properties of Alkali Activated Binders

Alkali-activated binders are promisors as an alternative to Portland cement and can be used in various applications due to their suitable properties such as mechanical strength, excellent durability, good resistance to acids or sulphate adhesion, and high fire resistance (Pacheco-Torgal, 2015; Provis, 2018). Several authors reported higher acid resistance of alkali-activated binders compared to Portland cement. The literature also shows reasonable resistance to citric acid, with minor changes in the materials and moderate changes when exposed to nitric and hydrochloric acids, especially over long periods of time (six months or more) (Bernal & Provis, 2014; Davidovits et al., 1990). Table 1 summarizes some reports that evaluated acid resistance, fire resistance and durability, and mechanical properties of alkali-activated binders and provides a brief account of the results.

One of the main advantages of alkali-activated binders reported in the literature is their high acid resistance, as their high alkalinity enables them to neutralize the hydronium ion (H_3O^+). Even under extreme acidic conditions, no significant degradation of alkali-activated materials has been observed. In general, acid attack is performed with nitric acid, sulfuric acid, hydrochloric acid, and acetic acid. In the study by Davidovits et al. (1990), an alkali-activated binder made from mine tailings (base metals, potash, coal, and uranium) was immersed in hydrochloric and sulfuric acids for four weeks, with mass loss reported by the authors to be between 6 and 7%. The same experiments were conducted with Portland cement concretes, with disastrous results, with mass loss as high as 95%. As reported by several authors in the literature, Portland cement generally exhibits higher mass losses than alkali-activated binders when exposed to acid attack, especially at low pH (Albitar et al., 2017; Davidovits et al., 1990; Fernandez-Jimenez et al., 2007; Pacheco-Torgal et al., 2012).

Table 1. Reports in the literature evaluating the resistance of alkali-activated binders

Precursor of alkali-activated binder	Highlights	Reference authors
Base metals, potash, coal, and uranium	Several alkali-activated binders were immersed in hydrochloric and sulfuric acid for 4 weeks. They observed that mass losses of between 6 and 7% occurred in the alkali-activated binders prior to this period. The same experiments were performed with Portland cement concretes, and the results were disastrous, with mass losses as 95%.	Davidovits et al. (1990)
Metakaolin mixtures (MK)	The stability of various binders prepared with MK and exposed to aggressive solutions such as seawater, sodium sulfate solution, and sulfuric acid for periods of 7 and 270 days was studied. All the solutions evaluated had little negative effect on the development of the microstructure and the strength of the materials studied, and a slight increase in their flexural strength with time (more than 90 days).	Palomo et al. (1999)
Metallurgy blast furnace slag	Freeze-thaw resistance was evaluated, and specimens were subjected to up to 300 freeze-thaw cycles, and mass and dynamic elasticity were evaluated. The relative dynamic elasticity was about 90%, even after 300 freeze-thaw cycles, resulting in excellent freeze-thaw resistance.	Fu et al. (2011)
Metakaolin (MK), ultrafine ground granulated blast furnace slag (UGGBFS), ceramic waste powder (CWP) and clay brick waste powder (CBWP)	Thermal resistance, durability, and fire behavior of lightweight mortars were evaluated for four different alkali-activated binders. The UGGBFS exhibited the highest electrical resistance and lowest water absorption, while CBWP with lightweight expanded clay aggregate (LECA) showed the worst in this evaluation. The best results in terms of thermal resistance were obtained for CWP and CBWP with LECA, which retained more than 49% of their original strength at the higher temperature evaluated.	Ameri et al. (2019)
Glass cullet	Three different glass cullet types were investigated: flat glass, hollow glass, and windshield glass under different concentrations of activating solution (KOH), curing times, and sulfate and acid attack (with magnesium sulfate) and acid attack (with sulfuric acid) were evaluated. Mass expansion (after 2 days) was observed in both attack experiments due to water intake into the pores. However, no significant mass expansion or degradation was observed, even before 218 days.	Idir et al. (2023)

Table 1. (Continued)

Precursor of alkali-activated binder	Highlights	Reference authors
Coal fly ashes	The effect of temperature on the alkali-activated binder made from coal fly ashes (with low or high Ca content) was investigated when cured at 80°C for 24 hours. It was found that exposure of the alkali-activated binder to temperatures above 700°C resulted in mold deformation and partial melting of the binder at 1000 °C. However, no major mineralogical changes were observed at 400°C. An increase in compressive strength was observed for all samples with low Ca content, while the effect was reversed for binders with high Ca content.	Seyedian Choubi & Meral Akgul (2022)
Slag-based concrete	The effect of temperature (from 25°C to 900°C) on the properties of alkali-activated slag-based concrete was studied. It was reported that and increase in compressive strength was observed at temperatures below 500°C, which was a significant improvement compared to Portland cement. In addition, the authors reported that the compressive strength of this alkali-activated binder could increase at high temperatures.	Palizi &Toufigh (2023)

Another advantage of alkali-activated binders over Portland cement is their thermal stability. The literature highlights that these binders have exceptional resistance to high temperatures and fire. In general, Portland concretes begin to decompose at temperatures above 300°C, while alkali-activated binders remain relatively stable even at 1000°C. This thermal property has been widely documented, suggesting that alkali-activated binders are suitable for buildings in areas of high fire risk. They show high performance at high temperatures and maintain structural integrity even under such conditions (Albitar et al., 2017; Palizi & Toufigh, 2023; Seyedian Choubi & Meral Akgul, 2022). In addition to their high temperature resistance, some authors reported excellent performance in freeze-thaw conditions. It is obvious that further studies on freeze-thaw resistance are required, nevertheless, the results published in the literature so far are promising (Albitar et al., 2017; Brooks et al., 2010; Fu et al., 2011).

The ability of alkali-activated binders to inhibit corrosion of steel reinforcement has also been investigated. Some reports in the literature demonstrate evidence that alkali-activated binders can prevent steel corrosion, which is one of the main factors negatively affecting the concrete structure of Portland concretes (Albitar et al., 2017). Chloride diffusion studies show that alkali-activated binders can reduce the risk of element diffusion, which can lead to steel corrosion (Albitar et al., 2017; Fernandez-Jimenez et al., 2007). There are studies showing that chloride diffusion in alkali-activated binders is lower than in Portland concretes, sometimes as much as half the diffusion rate (Provis et al., 2015).

Lloyd et al. (2010) demonstrated that the presence of calcium in alkali-activated binders is critical in reducing steel reinforcement. However, some studies show similar resistance to Portland concretes depending on binder content. To date, there are few reports on the abilities of alkali-activated binders to prevent corrosion of reinforced steel, and further research is needed (Fernandez-Jimenez et al., 2007; Lloyd et al., 2010; Monticelli et al., 2016; Provis et al., 2015).

Use of Wastes for Alkali-Activated Binders Production in Geotechnics: Potentials and Challenges

Alkali-activated binders are prepared by reacting a precursor with an activator under alkaline conditions, resulting in the formation of a cement gel such as

C-A-S-H or N(K)-A-S-H. These precursors serve as a source of aluminosilicate in the reaction and may have a high or low calcium content in their composition. An aluminosilicate source is required to produce the alkali-activated binders, which can be obtained from natural resources or from industrial by-products or wastes. Despite of the possibility of using natural resources such as metakaolin (Palomo et al., 1999), clay, feldspar or other minerals composed of an aluminosilicate as precursors, there is growing interest in the production of new cements using waste materials that are not normally used by industry in its products and are generally considered to be environmentally problematic. This interest is extremely important, since the recycling of materials considered as waste is one of the most important ways to conserve natural resources is (Garcia-Lodeiro et al., 2015; Miller & Myers, 2020; Pacheco-Torgal, 2015; Provis, 2018). Different precursors and calcium sources have been used for alkali-activation, such as: fly ash from thermal power plants (Hanjitsuwan et al., 2018; Seyedian Choubi & Meral Akgul, 2022); mining tailings such as tailings from a cooper/nickel mine, from a copper/zinc mine, from a platinum mine and from an iron ore mine, for exemple (Obenaus-Emler et al., 2020; Zhang et al., 2022); blast furnace slag (Amer et al., 2021; Palizi & Toufigh, 2023); glass waste (Idir et al., 2020); incinerator ash (Choeycharoen et al., 2022); carbide lime (Bruschi et al., 2021, 2022; Santos et al., 2022), rice husk ash (Pelissaro et al., 2023), sugarcane bagasse ash (see Chapter 5) (Ferrazzo et al., 2023; Tonini de Araújo et al., 2023); red ceramic waste (see Chapter 7) (Lotero et al., 2023); eggshell lime (Ferreira et al., 2023) and among others.

In addition, utilizing aluminosilicate waste produced near the point of use helps decrease the environmental and economic impacts linked with material transportation. This practice also enhances the competitiveness of alkali-activated binders, as suppliers of Portland cement are readily accessible. Despite the environmental interest, there is a commercial aspect that should be considered: the valorization of a waste that previously had no commercial value. Therefore, it is more interesting to develop new alkali-activated binders with industrial waste than to use natural resources (Garcia-Lodeiro et al., 2015; Miller & Myers, 2020; Pacheco-Torgal, 2015; Provis, 2018). In addition, the use of aluminosilicate waste generated close to the point of use helps reduce the environmental and economic impacts associated with the material transportation and increases the competitiveness of activated-alkali binders because Portland cement suppliers are easily accessible (Sandanayake et al., 2018; Zahmak et al., 2021).

In addition to the use of waste as precursor materials, alternative alkaline activators have been explored in alkali-activation, as commercial activators are responsible for severe environmental impacts and high costs in alkaline cements production and application (Zahmak et al., 2021). Alkaline activators produced from waste can be interesting alternatives to commercial activators (e.g., sodium hydroxide and sodium silicate) and can reduce the consumption of electricity, the use of new raw materials, and the production costs for activators (Adesanya et al., 2021). Soda waste from Na_2CO_3 production (Lin et al., 2020), steel desulfurization powder (Adesanya et al., 2020), waste glass (Abdulkareem et al., 2021), olive biomass ash, almond shell ash (Soriano et al., 2022), rice husk ash (Pelissaro et al., 2023), and sand tailings from the flotation process (Saldanha et al., 2023) are examples of wastes used in new alkaline activators. In addition, wastewater with high alkali content arc attractive options for alkaline activators in terms of cost and environmental impact, since they are liquid wastewaters that do not require treatment for use in alkali activation, such as: Bayer liquid from bauxite-alumina extraction in the Bayer process (Jamieson et al., 2015), and wastewater from aluminum extrusion (Cristelo et al., 2021) and the paper and pulp industry (Novo et al., 2022). Sustainable sources of alkaline activators, lower activator dosages, and sustainable energy sources for alkaline activator production are also important strategies for improving the environmental performance of alkaline cements in stabilizing of geotechnical materials (Zahmak et al., 2021).

In the geotechnical context, research is being conducted on the stabilization of various materials with alkali-activated binders or geopolymers, including: clayey (Corrêa-Silva et al., 2020; Coudert et al., 2019; Cristelo et al., 2021; Zahmak et al., 2021), expansive (Samuel et al., 2020; Syed et al., 2020; Tonini de Araújo et al., 2023), and sandy soils (Consoli et al., 2021); metal-contaminated soils (Komaei et al., 2023; Pu et al., 2022); bauxite (Bruschi et al., 2021, 2022), gold (Li et al., 2019; Pan et al., 2022; Santos et al., 2022), and iron mining tailings (Levandoski et al., 2023; Saldanha et al., 2023; Servi et al., 2022); reclaimed asphalt pavement (Goldoni et al., 2023; Pelissaro et al., 2023); waste foundry sand (Ferrazzo et al., 2023), and others (see Chapters 8, 9, 10, and 11). These studies primarily focus on investigating mechanical performance (e.g., strength, stiffness, and durability) and microstructure (e.g., mineralogy, chemical composition, and morphology). In some cases, the material stabilized with alkali-activated binder demonstrated competitive mechanical performance compared to the use of high-early strength Portland cement (Bruschi et al., 2021; Ferrazzo et al., 2023; Santos et al., 2022; Tonini de Araújo et al., 2023).

Recently, there has been a growing interest in evaluating metal leaching from matrices cemented with alkali-activated binders, which is critical given the use of waste materials that may contain contaminants (e.g., heavy metals) in their chemical composition (Bruschi et al., 2022; Ferrazzo et al., 2023; Komaei et al., 2023; Pelissaro et al., 2023; Pu et al., 2022). Leaching assessment therefore allows estimation of the long-term environmental risks that these materials may pose under different disposal and use scenarios (see Chapter 11).

Another emerging area of research is the evaluation of sustainability through life-cycle assessments of stabilization of soils or other geotechnical materials with alkaline cements (Saldanha, et al., 2023; Samuel et al., 2020). Life cycle assessments in conjunction with mechanical and leaching tests are critical for determining the technical feasibility and sustainability of the binder, and thus for selecting the one that meets the requirements for the intended application. However, sustainability assessment of material stabilization with traditional cements is a more recent development, especially when it comes to the use of alkali-activated binders and the three pillars (environmental, economic, and social). Therefore, the use of sustainability assessment tools for geotechnical materials stabilized with alkali-activated binders is a broad field of research to be explored.

Final Considerations

In this chapter, an overview of alkali-activated systems, the main properties of alkali-activated binders, the use of wastes in alkaline activation, and their applications in the geotechnical literature was given. Based on the nature of the cementitious components, alkali-activated binders are classified as high-calcium cements, low-calcium cements, and hybrid alkaline cements. In these different systems, the concentrations of the alkaline activator, the curing temperature, and the reactive products (e.g., N-A-S-H, C-A-S-H, C(N)-A-S-H gels) vary. Alkali-activated binders are characterized by properties such as chemical resistance to acids and sulphates, resistance to high temperatures, freeze-thaw, resistance, and the possibility of using a wide range of precursors compared to Portland cement.

Early studies developed alkali-activated binders based on metakaolin, slag, and fly ash. Over time, various industrial and agro-industrial wastes have been explored as precursor materials, calcium sources, or as alkaline activators. In geotechnical engineering, these alkali-activated binders are used to stabilize natural soils, contaminated soils, mining tailings, reclaimed asphalt pavement, waste foundry sand, etc. Literature studies focus on evaluating the mechanical behavior and microstructure of the cemented materials. Leaching of contaminants and the sustainability of stabilization of geotechnical materials with alkali-activated binders have gained increasing attention in recent years, but they are still relatively unexplored. Future research is expected to develop alkali-activated binders with residues that have not yet been used and to evaluate the mechanical behavior, leaching, and environmental, economic, and social impacts.

References

Abdulkareem, M., Havukainen, J., Nuortila-Jokinen, J. & Horttanainen, M. (2021). Environmental and economic perspective of waste-derived activators on alkali-activated mortars. *Journal of Cleaner Production*, *280*. https://doi.org/10.1016/j.jclepro.2020.124651.

Adesanya, E., Ohenoja, K., Di Maria, A., Kinnunen, P. & Illikainen, M. (2020). Alternative alkali-activator from steel-making waste for one-part alkali-activated slag. *Journal of Cleaner Production*, *274*, 123020. https://doi.org/10.1016/j.jclepro.2020.123020.

Adesanya, E., Perumal, P., Luukkonen, T., Yliniemi, J., Ohenoja, K., Kinnunen, P. & Illikainen, M. (2021). Opportunities to improve sustainability of alkali-activated materials: A review of side-stream based activators. *Journal of Cleaner Production*, *286,* 125558. https://doi.org/10.1016/j.jclepro.2020.125558.

Albitar, M., Mohamed Ali, M. S., Visintin, P. & Drechsler, M. (2017). Durability evaluation of geopolymer and conventional concretes. *Construction and Building Materials*, *136*, 374–385. https://doi.org/10.1016/j.conbuildmat.2017.01.056.

Amer, I., Kohail, M., El-Feky, M. S., Rashad, A. & Khalaf, M. A. (2021). Characterization of alkali-activated hybrid slag/cement concrete. *Ain Shams Engineering Journal*, *12*(1), 135–144. https://doi.org/10.1016/j.asej.2020.08.003.

Belaïd, F. (2022). How does concrete and cement industry transformation contribute to mitigating climate change challenges? *Resources, Conservation and Recycling Advances*, *15,* 200084. https://doi.org/10.1016/j.rcradv.2022.200084.

Benhelal, E., Shamsaei, E. & Rashid, M. I. (2021). Challenges against CO_2 abatement strategies in cement industry: A review. *Journal of Environmental Sciences, 104*, 84–101. https://doi.org/10.1016/j.jes.2020.11.020.

Bernal, S. A. & Provis, J. L. (2014). Durability of alkali-activated materials: Progress and perspectives. *Journal of the American Ceramic Society*, *97*(4), 997–1008. https://doi.org/10.1111/jace.12831.

Brooks, R., Bahadory, M., Tovia, F. & Rostami, H. (2010). Properties of alkali-activated fly ash: High performance to lightweight. *International Journal of Sustainable Engineering*, *3*(3), 211–218. https://doi.org/10.1080/19397038.2010.487162.

Bruschi, G. J., dos Santos, C. P., Levandoski, W. M. K., Ferrazzo, S. T., Korf, E. P., Saldanha, R. B. & Consoli, N. C. (2022). Leaching assessment of cemented bauxite tailings through wetting and drying cycles of durability test. *Environmental Science and Pollution Research*, *29*(39), 59247–59262. https://doi.org/10.1007/s11356-022-20031-5.

Bruschi, G. J., Pereira dos Santos, C., Tonini de Araújo, M., Tonatto Ferrazzo, S., Filipe Veloso Marques, S. & Consoli, N. C. (2021). Green stabilization of bauxite tailings: mechanical study on alkali-activated materials. *Journal of Materials in Civil Engineering, 33(11)*, 06021007. https://doi.org/10.1061/(ASCE)MT.1943-5533.0003949.

Choeycharoen, P., Sornlar, W. & Wannagon, A. (2022). A sustainable bottom ash-based alkali-activated materials and geopolymers synthesized by using activator solutions from industrial wastes. *Journal of Building Engineering*, *54*, 104659. https://doi.org/10.1016/j.jobe.2022.104659.

Consoli, N. C., Daassi-Gli, C. A. P., Ruver, C. A., Lotero, A., Scheuermann Filho, H. C., Moncaleano, C. J. & Lourenço, D. E. (2021). Lime–ground glass–sodium hydroxide as an enhanced sustainable binder stabilizing silica sand. *Journal of Geotechnical and Geoenvironmental Engineering*, *147*(10). https://doi.org/10.1061/(asce)gt.1943-5606.0002624.

Corrêa-Silva, M., Miranda, T., Rouainia, M., Araújo, N., Glendinning, S. & Cristelo, N. (2020). Geomechanical behaviour of a soft soil stabilised with alkali-activated blast-furnace slags. *Journal of Cleaner Production*, *267*, 122017. https://doi.org/10.1016/j.jclepro.2020.122017.

Coudert, E., Paris, M., Deneele, D., Russo, G. & Tarantino, A. (2019). Use of alkali activated high-calcium fly ash binder for kaolin clay soil stabilisation: Physicochemical evolution. *Construction and Building Materials*, *201*, 539–552. https://doi.org/10.1016/j.conbuildmat.2018.12.188.

Cristelo, N., Rivera, J., Miranda, T. & Fernández-Jiménez, A. (2021). Stabilisation of a plastic soil with alkali activated cements developed from industrial wastes. *Sustainability, 13*(8). https://doi.org/10.3390/su13084501.

Davidovits, J., Comrie, D., Paterson, J. & Ritcey, D. (1990). *Geopolymeric concretes for environmental protection*. *12*, 30–39.

EPA. (2023). *Global CO_2 emissions rebound in 2021 after temporary reduction during COVID19 lockdown*. https://joint-research-centre.ec.europa.eu/jrc-news-and-updates/global-co2-emissions-rebound-2021-after-temporary-reduction-during-covid19-lockdown-2022-10-14_en.

Fernandez Pales, A. & Leung, Y. (2018). *Technology Roadmap - Low-Carbon Transition in the Cement Industry - International Energy Agency*. 1–66. www.wbcsdcement.org.

Fernandez-Jimenez, A., García-Lodeiro, I. & Palomo, A. (2007). Durability of alkali-activated fly ash cementitious materials. *Journal of Materials Science*, *42*(9), 3055–3065. https://doi.org/10.1007/s10853-006-0584-8.

Ferrazzo, S. T., Araújo, M. T. de, Bruschi, G. J., Chaves, H. M., Korf, E. P. & Consoli, N. C. (2023). Mechanical and environmental behavior of waste foundry sand stabilized with alkali-activated sugar cane bagasse ash-eggshell lime binder. *Construction and Building Materials*, *383*, 131313. https://doi.org/10.1016/j.conbuildmat.2023.131313.

Ferreira, F. A., Desir, J. M., Lima, G. E. S. de, Pedroti, L. G., Franco de Carvalho, J. M., Lotero, A. & Consoli, N. C. (2023). Evaluation of mechanical and microstructural properties of eggshell lime/rice husk ash alkali-activated cement. *Construction and Building Materials*, *364*, 129931. https://doi.org/10.1016/j.conbuildmat.2022.129931.

Fu, Y., Cai, L. & Yonggen, W. (2011). Freeze-thaw cycle test and damage mechanics models of alkali-activated slag concrete. *Construction and Building Materials*, *25*(7), 3144–3148. https://doi.org/10.1016/j.conbuildmat.2010.12.006.

García-Lodeiro, I., Maltseva, O., Palomo, A. & Fernández-Jiménez, A. (2012). Cimenturi hibride alcaline: Partea I. Fundamente hybrid alkaline cements: Part I. Fundamentals. *Revista Română de Materiale/Romanian Journal of Materials*, 42(4), 330–335.

Garcia-Lodeiro, I., Palomo, A. & Fernández-Jiménez, A. (2015). An overview of the chemistry of alkali-activated cement-based binders. In *Handbook of Alkali-Activated Cements, Mortars and Concretes* (pp. 19–47). Elsevier Inc. https://doi.org/10.1533/9781782422884.1.19.

Garcia-Lodeiro, I., Palomo, A., Fernández-Jiménez, A. & MacPhee, D. E. (2011). Compatibility studies between N-A-S-H and C-A-S-H gels. Study in the ternary diagram Na_2O-CaO-Al_2O_3-SiO_2-H_2O. *Cement and Concrete Research, 41(9),* 923–931. https://doi.org/10.1016/j.cemconres.2011.05.006.

Goldoni, A. G., Pelissaro, D. T., Silveira, E., Prietto, P. D. M. & Rosa, F. D. (2023). Durability and mechanical long-term performance of reclaimed asphalt pavement stabilized by alkali-activation. *Soils and Rocks*, *46*(1), e2023007422. https://doi.org/10.28927/SR.2023.007422.

Hanjitsuwan, S., Phoo-ngernkham, T., Li, L. yuan, Damrongwiriyanupap, N. & Chindaprasirt, P. (2018). Strength development and durability of alkali-activated fly ash mortar with calcium carbide residue as additive. *Construction and Building Materials*, *162*, 714–723. https://doi.org/10.1016/j.conbuildmat.2017.12.034.

Harrisson, A. M. (2019). Constitution and specification of Portland cement. In *Lea's Chemistry of Cement and Concrete* (pp. 87–155). Elsevier. https://doi.org/10.1016/B978-0-08-100773-0.00004-6.

Idir, R., Cyr, M. & Pavoine, A. (2020). Investigations on the durability of alkali-activated recycled glass. *Construction and Building Materials*, *236*, 117477. https://doi.org/10.1016/j.conbuildmat.2019.117477.

Jamieson, E., McLellan, B., Van Riessen, A. & Nikraz, H. (2015). Comparison of embodied energies of Ordinary Portland Cement with Bayer-derived geopolymer products. *Journal of Cleaner Production*, *99*, 112–118. https://doi.org/10.1016/j.jclepro.2015.03.008.

Komaei, A., Noorzad, A. & Ghadir, P. (2023). Stabilization and solidification of arsenic contaminated silty sand using alkaline activated slag. *Journal of Environmental Management*, *344*, 118395. https://doi.org/10.1016/j.jenvman.2023.118395.

Le Quéré, C., Jackson, R. B., Jones, M. W., Smith, A. J. P., Abernethy, S., Andrew, R. M., De-Gol, A. J., Willis, D. R., Shan, Y., Canadell, J. G., Friedlingstein, P., Creutzig, F. & Peters, G. P. (2020). Temporary reduction in daily global CO_2 emissions during the COVID-19 forced confinement. *Nature Climate Change*, *10*(7), 647–653. https://doi.org/10.1038/s41558-020-0797-x.

Levandoski, W. M. K., Ferrazzo, S. T., Bruschi, G. J., Consoli, N. C. & Korf, E. P. (2023). Mechanical and microstructural properties of iron mining tailings stabilized with alkali-activated binder produced from agro-industrial wastes. *Scientific Reports*, *13*(1), 15754. https://doi.org/10.1038/s41598-023-42999-x.

Li, Y. Y., Zhang, T. T., Jia, S. B., Liu, J., Quan, X. H. & Zheng, W. (2019). Mechanical properties and leaching characteristics of geopolymer-solidified/stabilized lead-contaminated soil. *Advances in Civil Engineering*, *2019*. https://doi.org/10.1155/2019/6015769.

Li, Z., Zhang, J. & Wang, M. (2020). Structure, reactivity, and mechanical properties of sustainable geopolymer material: A reactive molecular dynamics study. *Frontiers in Materials*, *7*. https://doi.org/10.3389/fmats.2020.00069.

Lin, Y., Xu, D. & Zhao, X. (2020). Effect of soda residue addition and its chemical composition on physical properties and hydration products of soda residue-activated slag cementitious materials. *Materials*, *13*(7). https://doi.org/10.3390/MA13071789.

Lloyd, R. R., Provis, J. L. & Van Deventer, J. S. J. (2010). Pore solution composition and alkali diffusion in inorganic polymer cement. *Cement and Concrete Research*, *40*(9), 1386–1392. https://doi.org/10.1016/j.cemconres.2010.04.008.

Lotero, A., Moncaleano, C. J. & Consoli, N. C. (2023). Alkali-activated red ceramic wastes-carbide lime blend: An alternative alkaline cement manufactured at room temperature. *Journal of Building Engineering*, *65*, 105663. https://doi.org/10.1016/j.jobe.2022.105663.

Marvila, M. T., Azevedo, A. R. G. de & Vieira, C. M. F. (2021). Reaction mechanisms of alkali-activated materials. *Revista IBRACON de Estruturas e Materiais*, *14*(3). https://doi.org/10.1590/s1983-41952021000300009.

McLellan, B. C., Williams, R. P., Lay, J., Van Riessen, A. & Corder, G. D. (2011). Costs and carbon emissions for geopolymer pastes in comparison to ordinary Portland cement. *Journal of Cleaner Production*, *19*(9–10), 1080–1090. https://doi.org/10.1016/j.jclepro.2011.02.010.

Miller, S. A. & Myers, R. J. (2020). Environmental impacts of alternative cement binders. *Environmental Science and Technology*, *54*(2), 677–686. https://doi.org/10.1021/acs.est.9b05550.

Mohapatra, A. K. & Pradhan, B. (2022). Hybrid alkali activated cements (HAACs) system: A state-of-the-art review on fresh, mechanical, and durability behaviour. *Construction and Building Materials, 361*, 129636. https://doi.org/10.1016/j.conbuildmat.2022.129636.

Monteiro, P. J. M., Miller, S. A. & Horvath, A. (2017). Towards sustainable concrete. *Nature Materials*, *16*(7), 698–699. https://doi.org/10.1038/nmat4930.

Monticelli, C., Natali, M. E., Balbo, A., Chiavari, C., Zanotto, F., Manzi, S. & Bignozzi, M. C. (2016). A study on the corrosion of reinforcing bars in alkali-activated fly ash mortars under wet and dry exposures to chloride solutions. *Cement and Concrete Research*, *87*, 53–63. https://doi.org/10.1016/j.cemconres.2016.05.010.

Morgado, A., Hugues, P. & Vass, T. (2023). *Cement - IEA*. https://www.iea.org/energy-system/industry/cement.

Novo, C. C., Senff, L., Seabra, M. P., Novais, R. M. & Labrincha, J. A. (2022). The role of an industrial alkaline wastewater in the alkali activation of biomass fly ash. *Applied Sciences*, *12*(7), 3612. https://doi.org/10.3390/app12073612.

Obenaus-Emler, R., Falah, M. & Illikainen, M. (2020). Assessment of mine tailings as precursors for alkali-activated materials for on-site applications. *Construction and Building Materials*, *246*, 118470. https://doi.org/10.1016/j.conbuildmat.2020.118470.

Pacheco-Torgal, F. (2015). Introduction to handbook of alkali-activated cements, mortars and concretes. In *Handbook of alkali-activated cements, mortars and concretes* (pp. 1–16). Elsevier Inc. https://doi.org/10.1533/9781782422884.1.

Pacheco-Torgal, F., Abdollahnejad, Z., Camões, A. F., Jamshidi, M. & Ding, Y. (2012). Durability of alkali-activated binders: A clear advantage over Portland cement or an unproven issue? *Construction and Building Materials, 30,* 400-405. https://doi.org/10.1016/j.conbuildmat.2011.12.017.

Palizi, S. & Toufigh, V. (2023). Fire-induced damage assessment of cementless alkali-activated slag-based concrete. *Construction and Building Materials*, *393*, 132002. https://doi.org/10.1016/j.conbuildmat.2023.132002.

Palomo, A., Blanco-Varela, M. T., Granizo, M. L., Puertas, F., Vazquez, T. & Grutzeck, M. W. (1999). Chemical stability of cementitious materials based on metakaolin. *Cement and Concrete Research, 29(7)*, 997-1004. https://doi.org/10.1016/S0008-8846(99)00074-5.

Pan, Z., Zhang, C., Li, Y. & Yang, C. (2022). Solidification/stabilization of gold ore tailings powder using sustainable waste-based composite geopolymer. *Engineering Geology*, *309*, 106793. https://doi.org/10.1016/j.enggeo.2022.106793.

Park, S., Moges, K. A., Wu, S. & Pyo, S. (2022). Characteristics of hybrid alkaline cement composites with high cement content: flash set and high compressive strength. *Journal of Materials Research and Technology*, *17*, 1582–1597. https://doi.org/10.1016/j.jmrt.2022.01.105.

Pelissaro, D. T., Bruschi, G. J., Pavan Korf, E. & Dalla Rosa, F. (2023). Rice husk ash as an alternative soluble silica source for alkali-activated metakaolin systems applied to recycled asphalt pavement stabilization. *Transportation Geotechnics*, *39*, 100940. https://doi.org/10.1016/j.trgeo.2023.100940.

Provis, J. L. (2018). Alkali-activated materials. *Cement and Concrete Research*, 114, 40–48. https://doi.org/10.1016/j.cemconres.2017.02.009.

Provis, J. L., Palomo, A. & Shi, C. (2015). Advances in understanding alkali-activated materials. *Cement and Concrete Research*, *78,* 110–125. https://doi.org/10.1016/j.cemconres.2015.04.013.

Pu, S., Duan, W., Zhu, Z., Wang, W., Zhang, C., Li, N., Jiang, P. & Wu, Z. (2022). Environmental behavior and engineering performance of self-developed silico-aluminophosphate geopolymer binder stabilized lead contaminated soil. *Journal of Cleaner Production*, *379*, 134808. https://doi.org/10.1016/j.jclepro.2022.134808.

Saldanha, R. B., Caicedo, A. M. L., de Araújo, M. T., Scheuermann Filho, H. C., Moncaleano, C. J., Silva, J. P. S. & Consoli, N. C. (2023). Potential use of iron ore tailings for binder production: A life cycle assessment. *Construction and Building Materials*, *365*, 130008. https://doi.org/10.1016/j.conbuildmat.2022.130008.

Samuel, R., Puppala, A. J. & Radovic, M. (2020). Sustainability benefits assessment of metakaolin-based geopolymer treatment of high plasticity clay. *Sustainability*, *12*(24), 1–14. https://doi.org/10.3390/su122410495.

Sandanayake, M., Gunasekara, C., Law, D., Zhang, G. & Setunge, S. (2018). Greenhouse gas emissions of different fly ash based geopolymer concretes in building construction. *Journal of Cleaner Production*, *204*, 399–408. https://doi.org/10.1016/j.jclepro.2018.08.311.

Santos, C. P., Bruschi, G. J., Mattos, J. R. G. & Consoli, N. C. (2022). Stabilization of gold mining tailings with alkali-activated carbide lime and sugarcane bagasse ash. *Transportation Geotechnics*, *32*, 100704. https://doi.org/10.1016/j.trgeo.2021.100704.

Servi, S., Lotero, A., Silva, J. P. S., Bastos, C. & Consoli, N. C. (2022). Mechanical response of filtered and compacted iron ore tailings with different cementing agents: Focus on tailings-binder mixtures disposal by stacking. *Construction and Building Materials*, *349*, 128770. https://doi.org/10.1016/j.conbuildmat.2022.128770.

Seyedian Choubi, S. & Meral Akgul, C. (2022). High temperature exposure of alkali-activated coal fly ashes. *Journal of Building Engineering*, *59*, 105081. https://doi.org/10.1016/j.jobe.2022.105081.

Shah, I. H., Miller, S. A., Jiang, D. & Myers, R. J. (2022). Cement substitution with secondary materials can reduce annual global CO_2 emissions by up to 1.3 gigatons. *Nature Communications*, *13*(1), 5758. https://doi.org/10.1038/s41467-022-33289-7.

Soriano, L., Font, A., Borrachero, M. V., Monzó, J. M., Payá, J. & Tashima, M. M. (2022). Biomass ashes to produce an alternative alkaline activator for alkali-activated cements. *Materials Letters*, *308*, 131198. https://doi.org/10.1016/j.matlet.2021.131198.

Syed, M., GuhaRay, A. & Kar, A. (2020). Stabilization of expansive clayey soil with alkali activated binders. *Geotechnical and Geological Engineering*, *38*(6), 6657–6677. https://doi.org/10.1007/s10706-020-01461-9.

Tonini de Araújo, M., Tonatto Ferrazzo, S., Mansur Chaves, H., Gravina da Rocha, C. & Cesar Consoli, N. (2023). Mechanical behavior, mineralogy, and microstructure of alkali-activated wastes-based binder for a clayey soil stabilization. *Construction and Building Materials*, *362*, 129757. https://doi.org/10.1016/j.conbuildmat.2022.129757.

Wan, H., Yuan, L. & Zhang, Y. (2020). Insight into the leaching of sodium alumino-silicate hydrate (N-A-S-H) gel: A molecular dynamics study. *Frontiers in Materials*, *7*. https://doi.org/10.3389/fmats.2020.00056.

Yusslee, E. & Beskhyroun, S. (2022). The potential of one-part alkali-activated materials (AAMs) as a concrete patch mortar. *Scientific Reports*, *12*(1), 15902. https://doi.org/10.1038/s41598-022-19830-0.

Zahmak, A., Abdallah, M., Jarah, B. & Arab, M. G. (2021). Environmental performance of alkali-activated binders for ground improvement. *Transportation Geotechnics*, *31*, 100631. https://doi.org/10.1016/j.trgeo.2021.100631.

Zhang, N., Hedayat, A., Figueroa, L., Steirer, K. X., Li, H., Bolaños Sosa, H. G., Huamani Bernal, R. P., Tupa, N., Morales, I. Y. & Canahua Loza, R. S. (2022). Experimental studies on the durability and leaching properties of alkali-activated tailings subjected to different environmental conditions. *Cement and Concrete Composites, 130,* 104531. https://doi.org/10.1016/j.cemconcomp.2022.104531.

Chapter 5

Alkali-Activated Materials Using Sugarcane Bagasse Ash: A Bibliometric Review

Rafaela Pollon[1,*]
Caroline Müller[1]
Deise Trevizan Pelissaro[2]
Pedro Domingos Marques Prietto[3]
and Eduardo Pavan Korf[1]
[1]Graduate Program in Environmental Science and Technology, Universidade Federal da Fronteira Sul, Erechim, RS, Brazil
[2]Department of Civil Engineering, Universidade Regional Integrada do Alto Uruguai e das Missões, Erechim, RS, Brazil
[3]Graduate Program in Civil and Environmental Engineering, Universidade de Passo Fundo, Passo Fundo, RS, Brazil

Abstract

Concern for the conservation of natural resources is increasing every year. Portland cement is responsible for 7% of human CO_2 emissions from raw material extraction to its production and consumes up to 100 kWh of energy per ton produced. This bring the question of materials that are less harmful to the environment and have similar performance characteristics to Portland cement. Among these materials, alkali-activated materials, obtained by reacting a solid aluminosilicate with an alkaline activator, stand out. Industrial and agro-industrial wastes, such as sugarcane bagasse ash (SCBA), can be used to produce alkali-

* Corresponding Author's Email: rafaelapollon17@hotmail.com.

In: Environmental Geotechnics
Editors: E. Pavan Korf, P. Domingos Marques Prietto, C. Müller et al.
ISBN: 979-8-89113-651-9

activated materials, as they contain large amounts of alumina and silica. Therefore, a discussion on the conditions for the development of an alkali-activated binder based on SCBA was based on a bibliometric literature review. This chapter describes the bibliometric study of the papers published in the Scopus and Web of Science databases on alkali-activated materials, which was used to analyze and discuss the variables evaluated. The most commonly used activator was NaOH and the curing temperature was at room temperature. In the published papers, SCBA is generally not used as the sole precursor, which highlights another gap in these studies, especially when relating the effects of SCBA as the sole precursor to lower molarities of alkaline activators to achieve lower environmental and economic impacts.

Keywords: alkali-activation, cementing agent, variables, bibliometric

Introduction

The conscious use of natural resources available on Earth is a current and indispensable issue to allow the development of future generations. In Brazil, it is estimated that the production of one ton of cement consumes up to 100 kWh of energy (Gartner & Macphee, 2011) and generates an average of 600 kg of CO_2 (Freire, 2016). Globally, cement production is responsible for 7% of CO_2 emissions (SNIC, 2019). Therefore, there is a growing demand for alternative materials that can be utilized as partial or complete substitutes for traditional cement and that will have a minimal impact on the environment.

Alkali-activated binders can be used to replace conventional Portland cement (Provis, 2018), which could reduce greenhouse gas emissions by 44 to 64% (McLellan et al., 2011). However, the extraction of alkali-activated materials requires the use of precursors, which can be obtained from various sources, such as industrial wastes: fly ash (Provis, 2018), rice husk ash (Pollon, 2022), blast furnace ash (Langaro et al., 2017), sugarcane bagasse ash (Le et al., 2022) and others.

In this chapter, SCBA is highlighted because Brazil provides an ideal environment for its cultivation, as the crop requires a tropical, hot and humid climate, as well as well-aerated, deep soils with good moisture retention and high fertility (Marin, 2022). Due to these characteristics, Brazil is the largest producer in the world, with 572.9 million tons in the 2022/2023 harvest (CONAB, 2022). Sugarcane is a raw material for the production of sugar and ethanol, and its by-products and residues can be used for cogeneration, the production of animal feed, and as fertilizer for crops (EMBRAPA, 2021). Sugarcane processing produces the residue bagasse (Duarte, 2020), from which SCBA is obtained that can be used to produce alkali-activated material (Provis, 2018).

The final destination of SCBA is often incorrect or even used in agriculture and construction. For the latter two cases, studies are needed to show the impact of residues on individual soil types and crops so that their use does not lead to social, economic, and environmental losses. In the case of applications in civil construction, its potential as a reactive material must be evaluated to increase strength without leaching pollutants (see Chapter 11).

The lack of standards for the use of SCBA in civil construction also requires studies that analyze the application potential, efficiency, and social, environmental, and economic impacts of this by product. Therefore, it is important to study alternatives to conventional cements so that two environmental and economic strategies can be combined: the reduction of CO_2 emissions in cement production and the appropriate disposal of SCBA waste.

Although some studies have already been carried out, the production of alkali-activated binders from the use of SCBA with alkaline activators is still poorly addressed in the literature due to the lack of studies addressing some specific aspects, such as the optimization of the best conditions and variables in the production process of these binders. To fill this knowledge gap, this chapter aims to clarify, through a bibliometric literature review, which variables and study trends are relevant for the development of an alkali-activated binder based on SCBA.

Portland Cement

In 2019, the world produced 4,168.00 million tons of cement, with China leading the production with 2,300.00 million tons. In 2019, Brazil recorded an apparent cement consumption of 56.6 million tons, corresponding to a per capita consumption of 261 kg/inhabitant/year, ranking eighth in the world (SNIC, 2020).

Portland cement is the main component of concrete, which is one of the most consumed materials in the world, providing society with safe, comfortable housing and reliable modern infrastructure (Huntzinger & Eatmon, 2009). Concrete can be defined as a mixture of hydraulic cement, aggregates, and water that may contain additives that modify or improve some of its properties. Portland cement is a dry, powdery material and is not itself a binder, but develops binding properties when mixed with water and hydrated. Cement is said to be hydraulic when the hydration products are stable in an aqueous environment (Mehta & Monteiro, 2014).

The use of additives in cement has been practiced in Brazil for more than five decades, using by-products from other industrial activities in the form of reuse, in addition to less polluting raw materials. This leads to a diversification of applications and specific properties of cement and also contributes to the reduction of emissions and fuel consumption. The main additives used include steel slag, fly ash, calcined clays, and limestone fillers, which enable environmental sustainability and conserve non-renewable natural resources by using by-products from other production processes (SNIC, 2019).

Cement production consists of heating a mixture of limestone, clay, and bauxite at temperatures between 1200°C and 1500°C, where calcium carbonate decomposes into calcium oxide, resulting in the emission of CO_2 and the production of clinker (Lamas et al., 2013). Clinker, the main component of cement, is responsible for the emission of 90% of the CO_2 produced during cement production. These emissions are mainly generated during the calcination/decarbonization of the raw material, which, in addition to the combustion of fuels in the kiln, includes treatment to remove water, CO_2, and other gasses associated with certain substances (ABCP, 2012; Rezende et al., 2017).

A ton of cement produced in Brazil emits an average of 600 kg of CO_2, lower than the 900 kg per ton emitted as a global average (Freire, 2016). The cement industry is responsible for 7% of total man-made CO_2 emissions worldwide (SNIC, 2019), and consumes up to 100 kWh per ton of cement produced (Gartner & Macphee, 2011). Therefore, due to concerns about the environmental damage caused, alternatives are currently being sought that can be used as partial or full substitutes for conventional cements with a lower environmental impact, in addition to additives.

Alkali-Activated Materials: Reaction Mechanisms, Variables for Analysis and Waste

A class of materials that can be used to replace traditional Portland cement are alkali-activated materials, very versatile and adaptable binders (Provis, 2018). These materials are obtained by alkaline activation, a generic term for the reaction of a solid aluminosilicate (called a precursor) under alkaline conditions (triggered by an alkaline activator) to produce a rigid ligand, and synthesized by activating a source of solid aluminosilicate with an alkali hydroxide or silicate solution (Provis, 2018). However, when these mixtures are added to Portland cement in smaller amounts, they do not fall within the definition of alkaline activation because the Portland cement itself is the main source of alkalinity (Provis, 2018).

The reaction mechanisms of alkali-activated materials can be divided into three different classes: high calcium binders (Bernal et al., 2014), low calcium binders (Provis et al, 2014), and mixed binders or intermediate calcium binders (Provis & Bernal, 2014). Binders with high calcium content include materials rich in silicon and calcium, such as blast furnace slag. In this category of binders, the materials are activated under moderately alkaline conditions to produce a calcium aluminosilicate hydrate gel known as C-A-S-H (Pacheco-Torgal et al., 2015).

The low calcium reaction mechanism occurs in materials rich in aluminum and silicon, such as fly ash (Kumar et al., 2007), and the reaction requires more aggressive alkaline properties (Provis et al., 2014), such as higher temperatures (Fernández-Jiménez et al, 2008). This class of ligands produces an amorphous alkaline aluminosilicate hydrate gel (N-A-S-H) or a zeolitic precursor that could hypothetically evolve into a zeolite (Palomo et al., 2005).

There is also a third mechanism in which the high calcium reaction is combined with the low calcium reaction, using materials composed of silicon, aluminum, and calcium. This mechanism is called mixed or intermediate calcium and represents the coexistence of C-A-S-H and sodium aluminum silicate hydrate (N-A-S-H) gels (Yip et al., 2008).

In the production of alkali-activated materials, several factors affect the manufacturing process and the final properties of these materials, such as curing temperature, reaction mechanism (high, low or medium calcium content) (Mangat & Lambert, 2016), curing time, $NaOH/SiO_2$ ratio (Manjunath & Narasimhan, 2020), activator molarity (Consoli et al., 2022), and SiO_2/Al_2O_3 ratio (Kovalchuk & Krienko, 2009; Provis & Rees, 2009), among others.

The variables are essentials to the formation and properties of alkali-activated materials, the appropriate adjustment of these variables allows the development of customized building materials with the desired properties, such as resistance and durability contribute to the sustainability of civil construction. However, as research increases, so does the number of materials that can be used, especially as precursors. Consequently, the number of variables that can be used in the dosage of activated alkaline materials also increases. Therefore, standardization methods must be developed for the combinations of variables and their ranges.

Some of the precursor materials used can be obtained locally, such as local volcanic or clay soils, or industrial by-products such as fly ash (Provis, 2018), rice husk ash (Pollon, 2022), blast furnace ash (Langaro et al., 2017), sugarcane bagasse ash (Le et al., 2022) and others. To provide the amount of calcium necessary for the reaction mechanisms that use this material, some residues can also be used as a calcium source, such as carbide lime (Lotero et al, 2021) and eggshell lime (Saldanha et al, 2021) (see Chapter 4).

Sugarcane Bagasse Ash

Sugarcane bagasse ash stands out because Brazil is the largest producer of this crop, which allows a high production of this residue. Bagasse, produced during the processing of sugarcane, represents about 30% of the ground sugarcane, with about 60 to 90% of this material being used in the plants themselves. Burning bagasse for energy produces about 25 kg of SCBA per ton (FAPESP, 1998; Duarte, 2020). SCBA can also contain up to 86.2% silicon dioxide (SiO_2) by mass (Molin Filho et al, 2019), a compound that can be used to generate alkali-activated material (Provis, 2018).

An example of SCBA composition is quartz, magnetite and hematite minerals; and silica (60.65%), iron (13.87%) and aluminum (5.76%) oxides (Ferrazzo et al, 2023). This waste has to be prepared before used in alkali-activated materials, the most commonly used preparation methods are ash milling, bagasse burning control, or even ash burning. In this situation, it is known that particle size and shape have a significant effect on the viability of Portland cement-based materials (Ríos-Parada, 2017).

Data Sources and Bibliometric Analysis

Based on the information previously described, a bibliometric review was performed to evaluate works related to alkali-activation with ash from sugarcane bagasse. For this purpose, with no restriction time, the keywords and Boolean operators "alkali activat*" OR "alkali-activat*" OR geopolymer, AND "sugarcane bagasse ash" OR SCBA, excluding (NOT) "review" were used in the Scopus and Web of Science databases, the keyword "review" was excluded because the focus is papers with evaluation of laboratory tests.

Forty-five papers were found on the Scopus platform and 35 papers on the Web of Science. The papers were previously analyzed using RStudio software (version 4.2.3), excluding 28 duplicate papers, resulting in a total of 52 papers on this topic, with no restriction on the publication period. The first year with publication were 2013, 2022 was the year with more papers, with a growth graphic along years (Figure 1). The main keywords in these papers are bagasse with 23 mentions, followed by compressive strength with 21, sugarcane bagasse with 19, and durability, geopolymers, inorganic polymers, and sugarcane bagasse ash with 13 mentions (Figure 2).

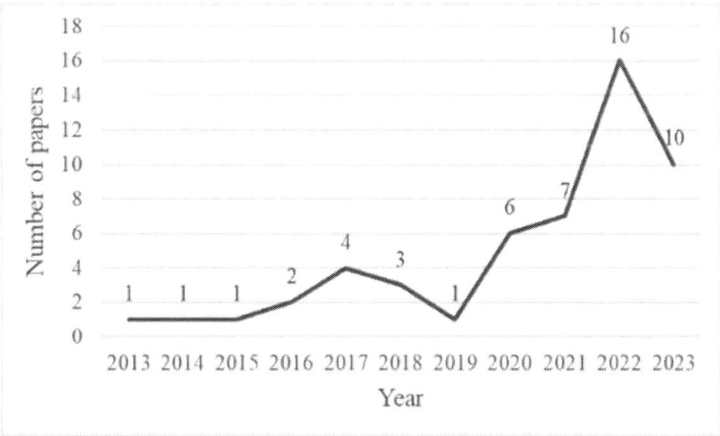

Figure 1. Annual Production.

Figure 2. Keyword cloud.

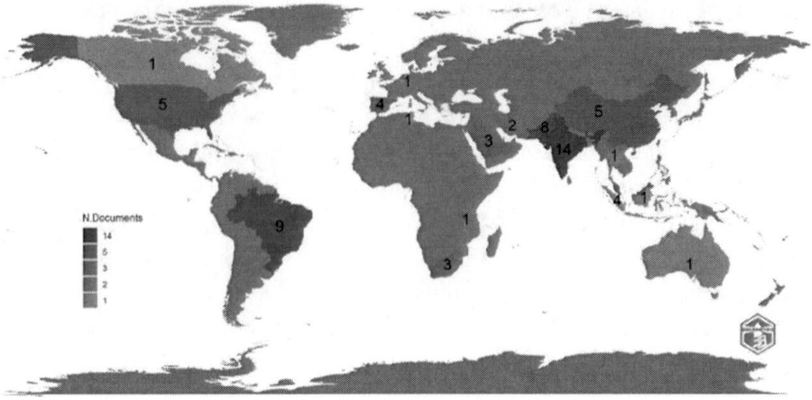

Figure 3. Country scientific production.

The countries with the largest scientific production on the topic of alkali-activation with SCBA are India, Brazil, Pakistan, China, and the United States of America, as shown in Figure 3. This can probably be explained by sugarcane production, which puts these countries in an advantageous position to exploit the properties of SCBA as a construction material. These countries, with their large populations and ever-growing infrastructure needs, are looking for sustainable solutions to meet construction needs and reduce environmental impacts.

Research and innovation using new building materials can play a fundamental role in development and promote the competitiveness and sustainability of these countries. Universities, research institutions, and industries in these countries have programs and funding focused on research in a variety of areas, including materials science and civil engineering. These investments help increase scientific production and publication of papers in peer-reviewed journals.

Although research is spread across different countries, there is still a small international collaboration network, as shown in Figure 4. Scientific collaboration and knowledge sharing can increase the quantity and quality of published scientific papers and promote progress in the field of alkali-activation with SCBA. Other factors may also influence this scenario, such as government policies, research priorities, and the interest of certain researchers and institutions in exploring this research area.

Figure 4. Collaboration network.

The Bibliometric Analysis

A bibliometric analysis of the papers was performed, following the exclusion criteria for papers with application of the ligand that did not focus exclusively on ligand development. Finally, 11 papers were selected to identify the main variables of interest and their impact on the alkali-activation process with SCBA for ligand development. Table 1 shows the selected papers and the variables evaluated: precursor, calcium source, alkaline activator, and curing time and temperature.

Araújo et al. (2023) studied strength, mineralogy, microstructure, and statistical analysis of alkali-activated SCBA – hydrated eggshell lime (HEL) pastes. The water/binder range used was between 0.7 and 1.2 and the SCBA/HEL range was 1.5, 2.33, 4.0, and 9.0.

França et al. (2023a) studied the reaction mechanisms in geopolymers prepared from SCBA. Sodium hydroxide with 97% purity and a sodium silicate composition of 14.98% Na_2O, 32.85% SiO_2, and 52.17% H_2O were used as activators. The molarity of the activators was 4 M, 8 M, and 12 M. The activator/precursor ratio was kept at 0.7, the relative humidity was above 65%, the weight of SCBA used was 400 g and 500 g, and that of Na_2SiO_3 was 80 g, 140 g, and 250 g, respectively. The weight of NaOH was 100 g, 140 g and 200 g. Molarities of 4 M, 8 M, and 12 M were used. The range of H_2O/Na_2O was from 6.68 to 20.37. The percentage of Na_2O was from 9.17 to

21.57. The silicate modulus (Ms or the range SiO_2/Na_2O) was from 0.31 to 1.67.

Aziz (2023) developed a geopolymer foamed concrete (GFC) containing sugarcane bagasse ash and fly ash: 100% recycled and cement-less concrete. Sodium hydroxide (NaOH) and sodium silicate (Na_2SiO_3) with a molarity of 12 M were used as activators. Intumescence was achieved with a 50% aqueous solution of hydrogen peroxide (H_2O_2). The ratio of alkali activators/solids was 0.25, while the ratio of sodium silicate to sodium hydroxide was kept constant at 2.5/1. Two percent hydrogen peroxide was used to produce a cellular or porous structure of the GFC. The fly ash (%)/bagasse ash (%) ratio ranged from 100/0 to 50/50.

França et al. (2023b) performed preliminary reactivity tests for precursors of alkali-activated materials. NaOH at a concentration of 8 mol/L served as a precursor. The range of H_2O/Na_2O was from 10.44 to 11.38 and SiO_2/Na_2O was 1.36. Thirty grams of precursor was always used (although the nature of the precursor was changed).

Mehmood et al. (2022) investigated the role of industrial by-products and metakaolin in the development of sustainable geopolymer mixtures: Upscaling from laboratory to pilot-scale. The alkaline activator was prepared by mixing 10 M NaOH solution with sodium silicate Na_2SiO_3 at a ratio (Na_2SiO_3/NaOH) of 1.5. The alkali percentage ranged from 0.53 to 0.6. The percentage of metakaolin varied from 10, 20, and 30%, fly ash was used at 70, 80, and 90%, and SCBA was used at 70, 80, and 90% of the total weight of the precursor. The ratio of alkaline activator to precursor ranged from 0.5 to 0.6.

Sheen & Le (2022) investigated an innovative use of sugarcane bagasse ash in green alkali-activated slag material: effects of activator concentration on blended pastes. A combination of 10 M sodium hydroxide (NaOH, 98% purity) and aqueous sodium metasilicate (9 wt% Na_2O, 28 wt% SiO_2, 63 wt% H_2O; SiO_2/Na_2O ratio of 3.11) was used as the activator, varying the silica modulus (Ms=SiO_2/Na_2O=0.6, 0.8, and 1.2) and sodium concentration (n=Na_2O/binder=6%, 8%, and 10%). A series of blends of GBFS/SCBA with mass ratios of 100/0, 90/10, 80/20, 70/30, and 60/40 were also prepared as starting materials. The total weight of the starting material (GBFS + SCBA) was kept at 1400 kg/m^3 for all blends. The liquid-to-solid ratio (L/S) was set at 0.35. The weight of the solid consists of SCBA, GBFS, and solid activators.

Table 1. Precursors, calcium source, activator, curing temperature and curing time of the selected papers

Precursors	Calcium source	Alkaline activator	Curing temperature (°C)	Curing time (days)	Reference
Sugarcane bagasse ash (SCBA)	Hydrated eggshell lime (HEL)	Sodium hydroxide (NaOH)	23	7	Araújo et al. (2023)
SCBA	No	NaOH and sodium silicate ($NaSiO_3$)	24	7 and 28	França et al. (2023a)
SCBA and fly ash (FA)	No	NaOH and Na_2SiO_3	Temperature-controlled environment	28	Aziz et al. (2023)
Silica fume (SF), granulated blast furnace slag (BFS), metakaolin (MK), mechanically treated SCBA, and mechanically and heat treated SCBA	No	NaOH	No information	28	França et al. (2023b)
MK, FA, and SCBA	No	NaOH and Na_2SiO_3	25	7, 28 and 56	Mehmood et al. (2022)
Ground blast furnace slag (GBFS) and SCBA	No	NaOH and Na_2SiO_3	Room temperature	3, 7, 14, and 28	Sheen & Le (2022)
High calcium lignite fly ash, and SCBA	High calcium lignite fly ash	NaOH and Na_2SiO_3	Room temperature	7 and 28	Chuewangkam et al. (2022)
BFS, MK, and SCBA	BFS	NaOH and Na_2SiO_3	23	7, 28 and 91	Sousa et al. (2022)
GBFS, SCBA	GBFS	NaOH	Room temperature	3, 7, 14, and 28	Le et al. (2022)
BFS, SCBA	BFS	Na_2SiO_3	No data	3 e 7	De Lima et al. (2021)
SCBA	Calcium aluminate cement	Na_2SiO_3	20, 40, 100, and 200	28	Pérez & García (2022)

Chuewangkam et al. (2022) investigated the mechanical and dielectric properties of composites of fly ash geopolymer and sugarcane bagasse ash. The 10 M NaOH solution was prepared by diluting NaOH flakes (98% purity) in deionized water, and the Na_2SiO_3 solution (12.53 wt% Na_2O, 30.24 wt% SiO_2, and 57.23 wt% H_2O) was used as the activator. The molar ratio of the Na_2SiO_3 solution was 2.49. The effect of adding SCBA (0, 10, 20, 30, and 40 wt%) to the fly ash geopolymers was investigated.

Sousa et al. (2022) studied the effects of non-calcined SCBA as an alternative precursor on the properties of alkali-activated pastes. A mixture of sodium hydroxide (NaOH, 8 mol/L) and sodium silicate solution (Na_2O 14.86%, SiO_2 32.07%, H_2O 52.16%) with an overall modulus (SiO_2/Na_2O molar ratio) of 1.43 was used as alkaline activators. The SiO_2/AlO_2 ratios used were 5.26, 5.77, 6.6, 6.85, 10.14, and 5.10. The GBFS/(GBFS + MK) ratios were 0, 0.6 and 1. The SCBA percentages were 0, 15, 30, 45, and 50. The CaO/SiO_2 ratios were 0.02, 0.28, 0.40, 0.45, and 0.55.

Le et al. (2022) investigated the potential use of SCBA for the development of alkali-activated materials. A 10M NaOH solution served as the activator. The silica modulus ($Ms=SiO_2/Na_2O$) was 0.6, 0.8, and 1.2. The liquid-to-solid (L/S) ratios were 0.35 and 0.4. The proportions of GBFS/SBA were 100/0, 90/10, 80/20, 70/30, and 60/40.

De Lima et al. (2021) studied the characterization of SCBA and its evaluation for use in alkali-activated slag mixtures. The sodium silicate was used with a silica modulus (SiO_2/Na_2O ratio) of 1.7. Two water/binder ratios were chosen: 0.28 and 0.40. After grinding the SCBA, replacement proportions of 10%, 20%, and 30% of the ground ash were used to evaluate flowability.

Pérez & García (2022) studied the effects of activation conditions on the compressive strength, microstructure, and reaction products of SCBA cements. Sodium silicate solutions with Na_2O concentrations of 8, 10, 12, and 14 wt%, based on the weight of SCBA, were used. In addition, the modulus of the sodium silicate solutions (SiO_2/Na_2O ratio) was changed at values of 0, 0.5, 1.0, and 1.5, using industrial-grade NaOH pellets. Na_2O (wt%) was 8, 10, 12, and 14, respectively.

It was found that in these studies, the most commonly used activator was sodium hydroxide, followed by sodium silicate. However, it was found that when sodium hydroxide is used in high concentrations to produce large quantities, it may require the use of special equipment due to its corrosive effect. In this sense, silicate activation can be considered better since it does not cause damage to equipment during handling, but for economic reasons, it is still not feasible compared to NaOH (Provis & Rees, 2009).

Although alkali activation can be performed under moderately alkaline conditions if a calcium source is used (Pacheco-Torgal et al., 2015), several authors have used activators with a concentration of 8 M or more, probably due to the high activation of aluminosilicates, leading to an increase in cost and environmental damage.

In most published works, room temperature was used to avoid additional heating costs and to make the application of these materials practical. Regarding curing times, there was a variation between 3 and 91 days, highlighting the periods of 7 and 28 days, which were evaluated in most studies and are considered traditional in evaluating the simple compressive strength (RCS) of traditional mortars and concretes. Ten of the eleven selected papers used SCBA together with another precursor or calcium source, making it difficult to analyze the properties and behavior of alkali-activated SCBA in isolation and leaving a gap regarding the actual effects of SCBA on the production of alkali-activated materials with and without calcium.

Final Considerations

The most commonly used activator has been sodium hydroxide, and when used in lower molarities, it can be considered for possible future applications, especially those involving large quantities of material, such as soil stabilization, mining tailings stabilization, sub-base of pavements or other civil construction applications.

A bibliometric analysis has shown that the variable "presence or absence of calcium" and the type of activator are very crucial factors for the behavior of alkali-activated materials. In terms of curing, it can be concluded that there is a tendency towards room temperature and curing times of 7 and 28 days. However, an important factor that needs to be further investigated is the molarity of the activator, since in most studies higher molarities than 8 M are used, resulting in mixtures with higher activator consumption and greater environmental impact.

Therefore, it is important that future research using SCBA as the sole starting material evaluate its characteristics and properties after activation. Room temperature curing also favors the performance of experiments and large-scale application of the material.

In addition to these factors, other variables were analyzed (alkali content, water/binder ratio, SiO_2/AlO_2 ratio, ratio between different precursors or activators, etc.), but it was not possible to identify a pattern in this respect. Another negative aspect is that there is no standardization of the representation of the material ratios: some authors use the material ratio in volume, others in mass, and still others in molar ratio, which makes it difficult to organize the available data and also affects the execution in future works.

References

ABCP – Associação Brasileira De Cimento Portland. (2012) *Relatório anual*. Brasília. Disponível em: http://arquivos.portaldaindustria.com.br/app/conteudo_24/2012/09/03/189/20121122172811763174i.pdf

Araújo, M. T., Ferrazzo, S. T., Bruschi, G. J., Silva, G. J. B. & Consoli, N. C. (2023). Strength, mineralogy, microstructure, and statistical analysis of alkali-activated sugarcane bagasse ash–eggshell lime pastes. *Journal of Materials in Civil Engineering*, *35*(6), 04023107. https://doi.org/10.1061/JMCEE7.MTENG-14539.

Aziz, I., Hassan, M. I. U., Haq, E. U. & Abbass, W. (2023). Development of geopolymer foam concrete incorporating sugarcane bagasse ash and fly ash; 100% recycled and cement-less concrete. *Arabian Journal for Science and Engineering*, *48*, 5655–5665. https://doi.org/10.1007/s13369-022-07511-z.

Bernal, S. A., Provis, J. L., Fernández-Jiménez, A., Krivenko, P. V., Kavalerova, E., Palacios, M. & Shi, C. (2014). Binder chemistry–high-calcium alkali-activated materials. *Alkali Activated Materials: State-of-the-Art Report, RILEM TC 224-AAM*, 59-91.

Chuewangkam, N., Nachaithong, T., Chanlek, N., Thongbai, P. & Pinitsoontorn, S. (2022). Mechanical and dielectric properties of fly ash geopolymer/sugarcane bagasse ash composites. *Polymers*, *14*(6), 1140. https://doi.org/10.3390/polym14061140.

CONAB – Companhia Nacional de Abastecimento (2022). *Nova estimativa de cana-de açúcar traz produção de 572,9 milhões de toneladas*. Brasília. https://www.conab.gov.br/ultimas-noticias/4725-nova-estimativa-de-cana-de-acucar-traz-producao-de-572-9-milhoes-toneladas#:~:text=De%20acordo%20com%20o%20levantamento,redu%C3%A7%C3%A3o%20de%202%2C2%25.

Consoli, N. C., Silvano, L. W., Lotero, A., Scheuermann Filho, H. C., Moncaleano, C. J. & Cristelo, N. (2022). Key parameters establishing alkali activation effects on stabilized rammed earth. *Construction and Building Materials*, *345*, 128299. https://doi.org/10.1016/j.conbuildmat.2022.128299.

De Lima, V. M. E., Barros, L. C. & de Melo Neto, A. A. (2021). Characterization of sugarcane bagasse ash (SBA) and its evaluation for use in alkali-activated slag mixtures. *Cerâmica, 67*, 123-130. https://doi.org/10.1590/0366-69132021673813038.

Duarte, J. B. (2020). *Estudo da adição conjunta da cinza do bagaço da cana-de-açúcar e do resíduo cerâmico na produção de tijolos maciços de solo-aglomerante.* Universidade Federal do Rio Grande do Norte.

EMBRAPA – Empresa Brasileira De Pesquisa Agropecuária (2022). *Cana - Abertura.* https://www.embrapa.br/agencia-de-informacao-tecnologica/cultivos/cana.

FAPESP. (1998). *Propriedades do bagaço da cana-de-açúcar.* https://revistapesquisa.fapesp.br/propriedades-do-bagaco-da-cana-de-acucar/.

Ferrazzo, S. T., de Araújo, M. T., Bruschi, G. J., Chaves, H. M., Korf, E. P. & Consoli, N. C. (2023). Mechanical and environmental behavior of waste foundry sand stabilized with alkali-activated sugar cane bagasse ash-eggshell lime binder. *Construction and Building Materials, 383*, 131313. https://doi.org/10.1016/j.conbuildmat.2023.131313.

Fernández-Jiménez, A., Monzó, M., Vicent, M., Barba, A. & Palomo, A. (2008). Alkaline activation of metakaolin–fly ash mixtures: Obtain of zeoceramics and zeocements. *Microporous and mesoporous materials, 108*(1-3), 41-49. https://doi.org/10.1016/j.micromeso.2007.03.024.

França, S., Figueiredo, P. F., Sousa, L. N., Silva, M. V. D. M. S., Borges, P. H. R. & da Silva Bezerra, A. C. (2023a). Reaction mechanisms in geopolymers produced from sugarcane bagasse ash. *Construction and Building Materials, 377*, 131124. https://doi.org/10.1016/j.conbuildmat.2023.131124.

França, S., Sousa, L. N., Silva, M. V. D. M. S., Borges, P. H. R. & Bezerra, A. C. D. S. (2023b). Preliminary reactivity test for precursors of alkali-activated materials. *Buildings, 13*(3), 693. https://doi.org/10.3390/buildings13030693.

Freire, D. (2016). Especialistas discutem papel da indústria do cimento nas emissões de CO_2. Fundação de Amparo à Pesquisa do Estado de São Paulo (FAPESP): São Paulo. https://agencia.fapesp.br/especialistas-discutem-papel-da-industria-do-cimento-nas-emissoes-de-co2/23449/.

Gartner, E. M. & Macphee, D. E. (2011). A physico-chemical basis for novel cementitious binders. *Cement and Concrete Research, 41*(7), 736-749. https://doi.org/10.1016/j.cemconres.2011.03.006.

Huntzinger, D. N. & Eatmon, T. D. (2009). A life-cycle assessment of Portland cement manufacturing: comparing the traditional process with alternative technologies. *Journal of cleaner production, 17*(7), 668-675. https://www.sciencedirect.com/science/article/pii/S0959652608000826.

Kovalchuk, G. & Krienko, P. V. (2009). Producing fire-and heat-resistant geopolymers. In *Geopolymers* (pp. 227-266). Woodhead Publishing. https://doi.org/10.1533/9781845696382.2.227.

Kumar, R., Kumar, S. & Mehrotra, S. P. (2007). Towards sustainable solutions for fly ash through mechanical activation. *Resources, Conservation and Recycling, 52*(2), 157-179. https://doi.org/10.1016/j.resconrec.2007.06.007.

Lamas, W. de Q., Palau, J. C. F. & De Camargo, J. R. (2013). Waste materials co-processing in cement industry: Ecological efficiency of waste reuse. *Renewable and Sustainable Energy Reviews, 19*, 200-207. https://doi.org/10.1016/j.rser.2012.11.015.

Langaro, E. A., Luz, C. A. D., Buth, I. S., Moraes, M. C. D., Pereira, J. I. & Matoski, A. (2017). A influência da composição química e da finura no desempenho de cimentos álcali ativados obtidos com escórias de alto forno. *Matéria*, *22*(1), 1-11. https://doi.org/10.1590/S1517-707620170001.0124.

Le, D. H., Sheen, Y. N. & Lam, M. N. T. (2022). Potential utilization of sugarcane bagasse ash for developing alkali-activated materials. *Journal of Sustainable Cement-Based Materials*, *11*(3), 199-209. https://doi.org/10.1080/21650373.2021.1920513.

Lotero, A., Consoli, N. C., Moncaleano, C. J., Neto, A. T. & Koester, E. (2021). Mechanical properties of alkali-activated ground waste glass–carbide lime blends for geotechnical uses. *Journal of Materials in Civil Engineering*, *33*(10), 04021284. https://doi.org/10.1061/(ASCE)MT.1943-5533.00039.

Mangat, P. & Lambert, P. (2016). Sustainability of alkali-activated cementitious materials and geopolymers. In *Sustainability of construction materials* (pp. 459-476). Woodhead Publishing. https://doi.org/10.1016/B978-0-08-100370-1.00018-4.

Manjunath, R. & Narasimhan, M. C. (2020). Alkali-activated concrete systems: A state of art. *New Materials in Civil Engineering*, *2020*, 459-491. https://doi.org/10.1016/B978-0-12-818961-0.00013-2.

Marin, F. R. (2022). Cana – Pré Produção – Características. Empresa Brasileira De Pesquisa Agropecuária (EMBRAPA). https://www.embrapa.br/agencia-de-informacao-tecnologica/cultivos/cana/pre-producao/caracteristicas.

McLellan, B. C., Williams, R. P., Lay, J., Van Riessen, A. & Corder, G. D. (2011). Costs and carbon emissions for geopolymer pastes in comparison to ordinary Portland cement. *Journal of Cleaner Production*, *19*(9-10), 1080-1090. https://doi.org/10.1016/j.jclepro.2011.02.010.

Mehmood, A., Irfan-ul-Hassan, M. & Yaseen, N. (2022). Role of industrial by-products and metakaolin in the development of sustainable geopolymer blends: Upscaling from laboratory-scale to pilot-scale. *Journal of Building Engineering*, *62*, 105279. https://doi.org/10.1016/j.jobe.2022.105279.

Mehta, P. K. & Monteiro, P. J. (2014). *Concreto. Microestrutura, propriedades e materiais* (3rd ed) McGraw Hill: Nova Iorque.

Molin Filho, R. G. D., Rosso, J. M., Volnistem, E. A., Vanderlei, R. D., Colpini, L. M. S., Ferrer, M. M., Paraíso, P. R., Jorge, L. M. M. (2019). Sugarcane bagasse ash micronized using air jet mills for green pozzolan in Brazil. *International Journal of Chemical Engineering*, *2019*, 5307098. https://doi.org/10.1155/2019/5307098.

Pacheco-Torgal, F., Labrincha, J., Leonelli, C., Palomo, A. & Chindaprasit, P. (2015). *Handbook of alkali-activated cements, mortars and concretes*. Elsevier.

Palomo, A., Fernández-Jimenez, A. & Kovalchuck, G. (2005). Some key factors affecting the alkali activation of fly ash. In *2nd International Symposium of Non-Traditional Cement and Concrete*.

Pérez, L. J. E. & García, Z. Z. Z. (2022). Influencia de las condiciones de activacion sobre la resistencia a la compresión, microestructura y productos de reacción de cementos de ceniza de bagazo de caña de azúcar. *Nexo Revista Científica*, 35(02), 476-486. https://doi.org/10.5377/nexo.v35i02.14624.

Pollon, R. (2022). *Estabilização de alto teor de material fresado com ligante álcali-ativado à base de metacaulim e cinza de casca de arroz*. Universidade Federal da Fronteira Sul.

Provis, J. L. (2018). Alkali-activated materials. *Cement and Concrete Research*, *114*, 40-48. https://doi.org/10.1016/j.cemconres.2017.02.009.

Provis, J. L. & Bernal, S. A. (2014). Binder chemistry–blended systems and intermediate Ca content. In *Alkali activated materials: State-of-the-art report, RILEM TC 224-AAM* (pp. 125-144). Dordrecht: Springer. https://doi.org/10.1007/978-94-007-7672-2_5.

Provis, J. L. & Rees, C. A. (2009). Geopolymer synthesis kinetics. In *Geopolymers* (pp. 118-136). Woodhead Publishing. https://doi.org/10.1533/9781845696382.1.118.

Provis, J. L., Fernández-Jiménez, A., Kamseu, E., Leonelli, C. & Palomo, A. (2014). Binder chemistry–low-calcium alkali-activated materials. In *Alkali Activated Materials: State-of-the-Art Report, RILEM TC 224-AAM*, (pp. 93-123). https://doi.org/10.1007/978-94-007-7672-2_4.

Rezende, M. F., Machado, F. C. S., Gouveia, A. M. C. D., Bezerra, A. C. D. S., Grillo, R. H. F. & Ortigara, Y. V. B. (2017). Substituição parcial do cimento Portland pela cinza de bagaço de cana-de-açúcar em habitações de interesse social. *Revista Agrogeoambiental*, *9*(1). https://doi.org/10.18406/2316-1817v9n12017914.

Ríos-Parada, V., Jiménez-Quero, V. G., Valdez-Tamez, P. L., & Montes-García, P. (2017). Characterization and use of an untreated Mexican sugarcane bagasse ash as supplementary material for the preparation of ternary concretes. *Construction and Building Materials, 157*, 83-95. https://doi.org/10.1016/j.conbuildmat.2017.09.060.

Saldanha, R. B., da Rocha, C. G., Caicedo, A. M. L. & Consoli, N. C. (2021). Technical and environmental performance of eggshell lime for soil stabilization. *Construction and Building Materials*, *298*, 123648. https://doi.org/10.1016/j.conbuildmat.2021.123648.

Sheen, Y. N. & Le, D. H. (2022). Innovative use of sugarcane bagasse ash in green alkali-activated slag material: Effects of activator concentration on the blended pastes. *Sugar Tech*, *24*(4), 1037-1051. https://doi.org/10.1007/s12355-022-01141-3.

SNIC – Sindicato Nacional da Indústria do Cimento (2019). Relatório Anual 2019. São Paulo. http://snic.org.br/assets/pdf/relatorio_anual/rel_anual_2019.pdf.

SNIC – Sindicato Nacional da Indústria do Cimento (2020). Relatório Anual 2020. São Paulo. http://snic.org.br/assets/pdf/relatorio_anual/rel_anual_2020.pdf.

Sousa, L. N., Figueiredo, P. F., França, S., de Moura Solar Silva, M. V., Borges, P. H. R. & Bezerra, A. C. D. S. (2022). Effect of non-calcined sugarcane bagasse ash as an alternative precursor on the properties of alkali-activated pastes. *Molecules*, *27*(4), 1185. https://doi.org/10.3390/molecules27041185.

Yip, C. K., Lukey, G. C., Provis, J. L. & Van Deventer, J. S. (2008). Effect of calcium silicate sources on geopolymerisation. *Cement and Concrete Research*, *38*(4), 554-564. https://doi.org/10.1016/j.cemconres.2007.11.001.

Chapter 6

Stabilization of Waste Foundry Sand with Alternative Binders

Claudio Luiz Pompermaier[1,*]
Suéllen Tonatto Ferrazzo[2]
Eduardo Pavan Korf[1]
and Pedro Domingos Marques Prietto[3]
[1]Graduate Program in Environmental Science and Technology,
Universidade Federal da Fronteira Sul, Erechim, RS, Brazil
[2]Graduate Program in Civil Engineering, Universidade Federal do Rio Grande do Sul,
Porto Alegre, RS, Brazil
[3]Graduate Program in Civil and Environmental Engineering,
Universidade de Passo Fundo, RS, Brazil

Abstract

Waste foundry sand (WFS) is the main waste generated in the metal foundry industry worldwide. It is estimated that for every ton of metal produced by smelting, about 0.6 tons of WFS is generated, making the annual global production of WFS about 62.64 million tons. This waste has been recycled only to a limited extent and is now largely sent to landfills, incurring economic and environmental costs. Several efforts have been made by the scientific community to find alternatives for the reuse of WFS. Because it has similar properties to natural sand, research has focused on civil construction and geotechnical engineering applications. WFS has been investigated as a substitute for natural and industrial sand or clay for various purposes, such as concrete, road base and subbase, filling road embankments and pipe beds, asphalt mixtures,

* Corresponding Author's Email: claudio.pompermaier@uffs.edu.br.

In: Environmental Geotechnics
Editors: E. Pavan Korf, P. Domingos Marques Prietto, C. Müller et al.
ISBN: 979-8-89113-651-9

ceramics, mortar and stabilization of clayey soils. In this context, stabilizing WFS with alternative cements or binders can provide an alternative to maximize the reuse of these wastes and minimize the impacts associated with landfill disposal and natural resource depletion. In addition, the use of alternative binders can reduce the impacts associated with the production of conventional binders such as Portland cement. The objective of this chapter is to provide an overview of the current situation of WFS, discuss its generation and disposal, and its main characteristics, with particular reference to research on the stabilization of WFS with alternative binders to support the assessment of this waste.

Keywords: waste valorization, alternative cements, geotechnics

Generation and Final Disposal of Waste Foundry Sand

Globally, the metal casting industry is of great importance due to the growing demand for metal parts and tools used primarily in the automotive industry (Dyer et al., 2018). Castings can be found in automobiles, trucks, agricultural machinery, mining and construction equipment, etc. The main metals cast are: iron, aluminium, magnesium, zinc, steel and copper alloys (American Foundry Society, 2020). In 2019, global foundry production was estimated at 109 million tons, with the largest producers and their shares being: China (44.5%), India (10.5%) and the United States (10.4%). Brazil occupies the ninth place with a share of 2.1% in this production (STATISTA, 2020).

In the production processes of the metallurgical industry, sand is used to make casting molds due to its properties such as good thermal conductivity and uniform particle size. Over time, this sand deteriorates due to heat and mechanical abrasion, making it impossible to return it to the production process and requiring appropriate final disposal, usually in landfills (Bhardwaj & Kumar, 2017; Tittarelli, 2018). WFS represents the largest share of by-products (Dyer et al., 2018; Tittarelli, 2018) and the largest pollutant (Iqbal et al., 2020) generated in the metal foundry industry. In Turkey, 300,000 tons of WFS are generated and landfilled annually (Bakis et al., 2006; Guney et al., 2010). In Brazil, more than 3 million tons of these wastes are generated annually (Mastella et al., 2014; Matos et al., 2019), mainly in the states of São Paulo, Santa Catarina, Rio Grande do Sul, and Minas Gerais. In other countries such as Spain, China, India, and the United States, an estimated 1 (Etxeberria et al., 2010), 48.8, 11.5, and 11.3 million tons of WFS are produced annually, respectively (Sabour et al., 2021).

In the USA, only 15% of waste is recycled (Torres et al., 2017), and the portion that is sent to landfills incurs an estimated annual cost of $135 million to $657 million (Siddique et al., 2018). A literature review found that there is limited and/or outdated data on WFS reuse in Brazil. In the state of Minas Gerais, 96.5% of 1.3 million tons of WFS were disposed of in industrial landfills in 2017 (81.25% by third parties and 15.25% self-managed) (FEAM, 2018). For Dyer et al. (2018), the disposal of WFS in landfills is still a problem, as large areas of these landfills become unproductive.

Characteristics of Waste Foundry Sand

There are two types of foundry sand, which differ according to the binder system used. Foundry sand usually consists of high-quality sand (80-95% quartz), a binder (usually bentonite, 4-10%), an additive (coal dust, 2-10%), and water (2-5%) and is referred to as "green sand" (Singh & Siddique, 2012). "Chemically bounded sand" has binders as a system: 1-3% phenolic, urethane, epoxy resins, and others. Green sand is generally the most used molding sand in the metal casting industry and accounts for more than 90% of the molding materials used (Siddique & Noumowe, 2008).

Among the WFS types, "green" sands are grayish, while chemically bonded sands are medium brown to light brown. In general, the particle size distribution is considered uniform, with 80-95% of the particles ranging from 0.6-0.15 mm (Siddique & Singh, 2011). The morphology of WFS is characterized by particles with sub-angular to round shapes, similar to natural sands such as Osorio sand (Ferrazzo et al., 2023). WFS is a rich source of silica, with a proportion ranging from 76 to 98%, a unit weight of grains ranging from 2.10 to 2.64 g.cm^{-3}, and a variable fine grain content (Table 1).

Leaching of WFS contaminants is another aspect that needs to be evaluated, as it is important for waste management and for the development of alternative uses (Iqbal et al., 2020; Siddique et al., 2010). Wastewater leachate may contain dissolved organic matter, inorganic macro components (e.g., iron, zinc, aluminium, sulfate, chloride, and ammonia), and heavy metals (e.g., lead, nickel, copper, and mercury) (Siddique et al., 2010). Penkaitis & Sígolo (2012) classified a sample of WFS as non-hazardous waste, according to NBR 10004 (ABNT, 2004). However, groundwater contamination was detected at the landfill where this waste is disposed. This contamination is due to leaching and solubilization of WFS associated with weathering and reactions with elements present in the landfill area. On the other hand,

Ferrazzo et al. (2023) stated that the studied WFS sample is classified as non-hazardous and inert waste (class II-B) according to NBR 10004 (ABNT, 2004). Therefore, understanding the leaching behavior of WFS is important for implementing waste reuse or recycling strategies.

Table 1. Physical and chemical properties of WFS samples (green sand)

Unit weight of grains (g.cm^{-3})	Fines (>0.075mm) (%)	Major chemical composition		Reference
		SiO_2 (%)	Al_2O_3 (%)	
-	> 6	96.73	0.59	Bakis et al. (2006)
2.45	24	98.00	0.80	Guney et al. (2006)
2.20	8	87.91	4.70	Siddique et al. (2009)
-	< 10	84.90	5.21	Etxeberria et al. (2010)
2.61	18	78.81	6.32	Siddique & Singh (2011)
2.55	0	76.00	4.45	Sahmaram et al. (2011)
-	0	85.40	3.64	Alonso-Santurde et al. (2012)
2.10	13.2	81.85	10.41	Basar & Aksoy (2012)
-	0	88.50	2.70	Navarro-Blasco et al. (2013)
-	0.3	87.91	4.70	Khatib et al. (2013)
2.61	0	83.93	0.02	Thaarrini & Ramasamy (2016)
2.59	2	84.15	11.82	Arulrajah et al. (2017)
2.45	9	94.10	1.70	Torres et al. (2017)
2.10	-	81.50	0.95	Siddique et al. (2018)
2.19	0	78.50	5.50	Bhardwaj & Kumar (2019)
2.64	0	84.90	5.21	Bhardwaj & Sharma (2020)
2.38	-	78.10	9.69	Matos et al. (2020)
-	-	86.60	2.91	Apithanyasai et al. (2020)
2.61	3.3	90.40	2.94	Ferrazzo et al. (2023)

Valorization of Waste Foundry Sand in Civil Engineering

In Brazil, government regulations and laws have been created to establish guidelines for the use of WFS in the following areas: asphalt concrete and non-structural concrete artefacts (ABNT, 2009; CETESB, 2007; Law No. 17,479, 2018; CONSEMA, 2013); daily covering material in landfills (ABNT, 2009; Law no. 17,479, 2018); production of tiles, bricks, and other fired clay articles for ceramic items, laying of pipes and artefacts for paving, base, subbase, subgrade reinforcement for the construction of roads, highways, and urban streets (Law No. 17,479, 2018). In addition, the United States Environmental Protection Agency (USEPA) has a guidance document with tools to encourage USA states to develop WFS reuse programs (USEPA, 2006).

Table 2. Synthesis of studies on WFS with alkaline activation

WFS function	Precursors and activators	Material developed	Proper replacement	Reference
Natural sand substitute	BFS, bottom ash, NaOH, Na_2SiO_3	Geopolymer concrete paver blocks	50%	Thaarrini & Ramasamy (2016)
Substitute for natural and industrial sands	FA, NaOH, Na_2SiO_3	Geopolymer concrete paver blocks	60%	Kewal (2017)
Precursor	WFS, NaOH, Na_2SiO_3	Geopolymer	-	Doğan-Sağlamtimur (2018)
Natural sand substitute	FA, BFS, NaOH, Na_2SiO_3	Geopolymeric and alkali-activated concretes	40%	Bhardwaj & Kumar (2019)
Filler material	FA, bentonite, NaOH, Na_2SiO_3	Geopolymer bricks	-	Philip et al. (2019)
Filler material substitute	FA, BFS, NaOH	Geopolymer	20%	Al-Mashhadani & Canpolat (2020)
Substitute for natural and industrial sands	FA, electric arc furnace slag, NaOH, Na_2SiO_3	Geopolymer blocks	30% WFS passing the #200	Apithanyasai et al. (2020)
Precursor	Desilicated foundry sand (DFS), NaOH	Geopolymer	-	Mashifana & Sithole (2020)
Natural sand substitute	Ground granulated blast-furnace (GGBF) slag NaOH, Na_2SiO_3	Geopolymer mortar	-	Sabour et al. (2022)
Natural sand substitute	Ground granulated BFS	Concrete	30%	Sithole et al. (2022)
Precursor	WFS, Aluminium slag, NaOH	Geopolymer	-	Sithole & Nkosi (2022)
Metakaolin substitute	Metakaolin, NaOH,Na_2SiO_3	Geopolymer	30%	Sgarlata et al. (2023)
Stabilized material	Sugarcane bagasse ash, eggshell lime, NaOH	WFS stabilized with alkali-activated binder	-	Ferrazzo et al. (2023)

Several efforts have been made by the scientific community to seek alternatives for WFS reuse, primarily through research aimed at its application in civil construction and geotechnical engineering. WFS has been investigated as a substitute for natural and industrial sand or clay for various purposes, such as concrete (Coppio et al., 2019; Iqbal et al., 2020; Siddique et al., 2018), road base and subbase (Moraes et al., 2023; Vinoth et al., 2022; Yazoghli-Marzouk et al., 2014; Zhang et al., 2021), the filling of embankments on highways and pipe beds (Arulrajah et al., 2017; Iqbal et al., 2019; Yaghoubi et al., 2020), as a hydraulic barrier (Osinubi & Moses, 2015), asphalt mixtures (Dyer et al., 2018), ceramics (Lin et al., 2018), mortars (Matos et al., 2020; Paul et al., 2022), and stabilization of clay soils (Bhardwaj et al, 2021; Kumar & Sharma, 2023; Patel & Chauhan, 2020). However, most research on the reuse of WFS is limited to its application in conventional concrete and its use as a partial replacement for natural sand (Bhardwaj & Kumar, 2019) or industrial sand. In other building materials such as ceramics and mortar, WFS is used as partial replacements of raw materials, limited to their content of fines and/or undergoing a calcination process (Lin et al., 2018; Matos et al., 2020; Paul et al., 2022).

In addition, in recent years, WFS has been explored in various alkali-activated materials in civil engineering, as shown in Table 2. However, in alkali-activated materials and geopolymers, the use of WFS is limited to additions of 20-60% and/or its percentage of fines (Al-mashhadani & Canpolat, 2020; Apithanyasai et al, 2020; Bhardwaj & Kumar, 2019).

In this context, stabilizing WFS with alternative cements or binders may be an alternative to maximize the reuse of this waste, minimize the impacts associated with landfill disposal and exploitation of natural resources, and contribute to making it a value-added material. According to Ferrazzo et al. (2023), stabilization of WFS with waste-derived binders is a form of contribution to the circular economy and can replace artificially cemented and compacted sand in geotechnical applications, such as pipe beds, shallow foundations, pavement bases and subbases, etc.

Waste Foundry Sand Stabilized with Alternative Binders

Recent geotechnical literature has investigated the stabilization of WFS with various alternative binders produced from industrial and/or agro-industrial wastes or where wastes are incorporated into conventional binders. One of the most important evaluations in the application of alternative binders for the stabilization of WFS is their mechanical performance compared to traditional binders such as Portland cement. Indeed, the different starting materials lead to different properties of the cementing gels and consequently of the cemented material.

In this context, Zhang et al. (2021a) investigated the stabilization of WFS for structural pavement applications using three types of alternative binders: coal fly ash, biomass fly ash (bioash), and an ecological commercial cement Ecolan (produced from a mixture of coal ash, bioash, lime, and cement). Unconfined compressive strength (UCS) results after 7 days of curing showed that biomass fly ash alone did not help stabilize WFS. However, with a small dose of calcium (in this case, 1% fast-setting Portland cement), the results were better than samples with coal fly ash, which also contained 1% Portland cement in the mixture. The behavior of the mixtures was less brittle than that of the reference specimens stabilized with Portland cement only. However, none of the specimens met the strength requirements necessary for the intended application: 3-5 MPa after 7 days.

As an alternative to improve the strength results of stabilized WFS without increasing the binder content or substituting the use of alternative binders, Zhang et al. (2021b) improved the particle size of the material by using coarser aggregates (crushed rock, 0-20 mm). The binders used were the same as in the previous study, and the specimens were tested after 7, 28, and 91 days of curing. The results showed that the change in granulometry allowed all specimens to meet the strength requirements (3-5 MPa) after 7 days. After 28 days, the mixtures stabilized with coal fly ash or bio ash did not meet the requirements for this curing time. However, after 91 days, there was almost no difference in resistance between mixtures with cement, fly ash or bio ash, proving that alternative binders can provide similar performance to conventional cement, considering their specificities. The ecological cement "Ecolan" obtained the highest resistance among all samples after 28 and 91 days of curing.

Another study on the stabilization of WFS with alternative binders based on residual materials was carried out by Pelisser et al. (2023). In this case, mixtures of rice husk ash-carbide lime and rice husk ash-commercial hydrated lime were used as binders. The results obtained showed an increase in the UCS of the specimens cured for 90 days. Ash content and dry unit weight were the factors that most affected the strength of the material. The strengths obtained after 90 days of curing were in the range of 1.5-2.0 MPa. WFS stabilized with rice husk ash- commercial hydrated lime had slightly higher strengths than the binder with carbide lime due to the excess of free Ca ions in the mixture with carbide lime.

In comparison, the highest strengths obtained by Zhang et al. (2021b) can be attributed to the improvement in granulometry assumed by the author, a higher specific weight, and the common use of a commercial binder (quick cement). However, regarding the reuse of WFS, Pelisser et al. (2023) used 100% of the residue, while Zhang et al. (2021b) used only 25% of the WFS and the other 75% consisted of crushed stone.

Alkaline activation of amorphous aluminosilicate waste sources is one way to improve resistance. Ferrazzo et al. (2023) stabilized WFS with an alkali-activated binder derived from sugarcane bagasse ash and eggshell lime activated with a sodium hydroxide solution. For comparison, stabilization was performed with the same binder dosages using high initial strength Portland cement (type III). The authors found that the WFS binder mixtures had the same or better mechanical performance than the WFS Portland cement mixtures, in terms of strength, initial stiffness, and durability. In addition, the formation of N-A-S-H gel was observed in the matrices cemented with the alkali-activated binder, justifying the satisfactory mechanical results.

Thus, from these studies, we can conclude that the mechanical strength of WFS stabilized with alternative binders varies depending on the materials used as stabilizers, the degree of compaction of the mixture, and the presence or absence of other materials to correct particle size. The application of different strategies, such as the improvement of granulometry, compaction, or the addition of alkaline activators, is able to improve the mechanical performance of cemented WFS, thus expanding its potential use as a geotechnical material. However, the importance of previous studies on the dosage of these binders is evident, as well as studies on the behavior of this type of material in the field.

Another problem with the use of alternative binders, especially when using wastes, is their leaching behavior. Various wastes may contain impurities with leaching potential, which may make their use as geotechnical materials impossible. Therefore, it is important to check the stabilization/solidification of WFS cemented with waste-based binders for contaminant inclusions (see Chapter 11).

Regarding this aspect, tests conducted by Ferrazzo et al. (2023) showed that WFS stabilized with an alkali-activated binder immobilized the metals (Cd, Cr, Mn, Pb, and Zn) contained in the waste used, making the composite safe for use in the field. A similar result was obtained by Pelisser et al. (2023), in which the cement matrix of rice husk ash and carbide lime also encapsulated the metals contained in WFS and other materials, thus showing no toxicity and consequently no environmental risks for use in engineering projects.

Finally, no less important is the question of the sustainability of alternative binders compared to conventional binders. Studies in this regard are still in the early stages. Ferrazzo et al. (2024) evaluated the three pillars of sustainability (environmental, economic, and social) of the alternative alkali-activated binder derived from sugarcane bagasse ash, eggshell lime, and sodium hydroxide used to stabilize WFS compared to stabilization with Portland cement at the same durability after 7 days. The result, presented using a sustainability index, showed that the alternative binder had the same score as Portland cement in environmental and economic terms, and received a better score in terms of social impact (Ferrazzo et al., 2024). The authors noted that the environmental impact of alkali-activated binders was mainly related to alkaline activators such as sodium hydroxide. The use of alternative activators and lower dosage can improve the environmental performance of alkaline cements (Zahmak et al., 2021).

In this sense, there is still a lack of studies comparing the environmental, economic, and social costs of disposing of various wastes in landfills relative to their use as materials for alternative binders, since non-renewable natural resources are used to produce Portland cement or commercial lime.

Final Considerations

This chapter presented the issues of WFS management and research related to the reuse of these wastes in engineering, focusing on geotechnical applications and highlighting key findings and trends. In general, the mechanical performance of WFS stabilized with alternative binders varies significantly depending on the properties of the binders used, the degree of compaction of the mixture, and the presence or absence of other materials to correct particle size. Therefore, dosage studies are essential for the use of alternative binders.

Alternative binders used to stabilize WFS can achieve similar mechanical performance to traditional binders such as Portland cement, provided that their specific features are considered, especially in terms of raw materials and curing time, which tends to be longer.

The studies carried out show that alternative binders are able to immobilize pollutants present in WFS and also in the waste used to produce the binders in their structure, making the material safe for applications as a geotechnical material.

In terms of sustainability, alternative binders can be more sustainable than traditional binders in a number of ways, which vary depending on the materials used. Therefore, life cycle assessments are critical in determining the most sustainable stabilization strategies for WFS. Finally, it is expected that the information discussed in this chapter can be used to develop new research and strategies for the sustainable reuse of WFS in geotechnical engineering.

References

Al-Mashhadani, M. M. & Canpolat, O. (2020). Effect of various NaOH molarities and various filling materials on the behavior of fly ash based geopolymer composites. *Construction and Building Materials*, *262*, 120560. https://doi.org/10.1016/j.conbuildmat.2020.120560.

Alonso-Santurde, R., Coz, A., Viguri, J. R. & Andrés, A. (2012). Recycling of foundry by-products in the ceramic industry: Green and core sand in clay bricks. *Construction and Building Materials*, *27*(1), 97–106. https://doi.org/10.1016/j.conbuildmat.2011.08.022.

American Foundry Society. (2020). *About metalcasting*. https://www.afsinc.org/about-metalcasting.

Apithanyasai, S., Supakata, N. & Papong, S. (2020). The potential of industrial waste: Using foundry sand with fly ash and electric arc furnace slag for geopolymer brick production. *Heliyon*, *6*(3), e03697. https://doi.org/10.1016/j.heliyon.2020.e03697.

Arulrajah, A., Yaghoubi, E., Imteaz, M. & Horpibulsuk, S. (2017). Recycled waste foundry sand as a sustainable subgrade fill and pipe-bedding construction material: Engineering and environmental evaluation. *Sustainable Cities and Society*, *28*, 343–349. https://doi.org/10.1016/j.scs.2016.10.009.

Associação Brasileira de Normas Técnicas – ABNT. (2004). *Resíduos sólidos - Classificação*. (ABNT NBR 10004: 2004). ABNT. https://www.normas.com.br/.

Associação Brasileira de Normas Técnicas – ABNT. (2009). *Areia descartada de fundição — Diretrizes para aplicação em asfalto e em aterro sanitário*. (ABNT NBR 15702: 2009). ABNT. https://www.normas.com.br/.

Bakis, R., Koyuncu, H. & Demirbas, A. (2006). An investigation of waste foundry sand in asphalt concrete mixtures. *Waste Management and Research*, *24*(3), 269–274. https://doi.org/10.1177/0734242X06064822.

Basar, H. M. & Deveci Aksoy, N. (2012). The effect of waste foundry sand (WFS) as partial replacement of sand on the mechanical, leaching and micro-structural characteristics of ready-mixed concrete. *Construction and Building Materials*, *35*, 508–515. https://doi.org/10.1016/j.conbuildmat.2012.04.078.

Bhardwaj, A. & Sharma, R. K. (2020). Effect of industrial wastes and lime on strength characteristics of clayey soil. *Journal of Engineering, Design and Technology*, *18*(6), 1749–1772. https://doi.org/10.1108/JEDT-12-2019-0350.

Bhardwaj, B. & Kumar, P. (2017). Waste foundry sand in concrete: A review. *Construction and Building Materials*, *156*, 661–674. https://doi.org/10.1016/j.conbuildmat.2017.09.010.

Bhardwaj, B. & Kumar, P. (2019). Comparative study of geopolymer and alkali activated slag concrete comprising waste foundry sand. *Construction and Building Materials*, *209*, 555–565. https://doi.org/10.1016/j.conbuildmat.2019.03.107.

Bhardwaj, A., Sharma, R. K. & Sharma, A. (2021). Stabilization of clayey soil using waste foundry sand and molasses. In: *Sustainable Development Through Engineering Innovations* (pp. 641–649). Springer Singapore.

CETESB – Environmental Sanitation Technology Company. (2007). *Procedimentos para gerenciamento de areia de fundição*. Companhia de Tecnologia de Saneamento Ambiental, Governo do Estado de São Paulo. https://cetesb.sp.gov.br/residuossolidos/wp-content/uploads/sites/26/2013/11/20100621111349_14c906d046b6f3c904a8e859375107c5.pdf.

Coppio, G. J. L., de Lima, M. G., Lencioni, J. W., Cividanes, L. S., Dyer, P. P. O. L. & Silva, S. A. (2019). Surface electrical resistivity and compressive strength of concrete with the use of waste foundry sand as aggregate. *Construction and Building Materials*, *212*, 514–521. https://doi.org/10.1016/j.conbuildmat.2019.03.297.

Doğan-Sağlamtimur, N. (2018). Waste Foundry Sand Usage for Building Material Production: A First Geopolymer Record in Material Reuse. *Advances in Civil Engineering*, *2018*. https://doi.org/10.1155/2018/1927135.

Dyer, P. P. O. L., de Lima, M. G., Klinsky, L. M. G., Silva, S. A. & Coppio, G. J. L. (2018). Environmental characterization of Foundry Waste Sand (WFS) in hot mix asphalt (HMA) mixtures. *Construction and Building Materials*, *171*, 474–484. https://doi.org/10.1016/j.conbuildmat.2018.03.151.

Etxeberria, M., Pacheco, C., Meneses, J. M. & Berridi, I. (2010). Properties of concrete using metallurgical industrial by-products as aggregates. *Construction and Building Materials*, *24*(9), 1594–1600. https://doi.org/10.1016/j.conbuildmat.2010.02.034.

FEAM – Fundação Estadual do Meio Ambiente (2018). *Inventário de resíduos sólidos industriais: Ano base 2017*. Fundação Estadual do Meio Ambiente. http://www.feam.br/images/stories/2018/RESIDUOS/Relat%C3%B3rio_Invent%C3%A1rio_Industria_2018_ano_base_2017.pdf.

Ferrazzo, S. T., Araújo, M. T. de, Bruschi, G. J., Chaves, H. M., Korf, E. P. & Consoli, N. C. (2023). Mechanical and environmental behavior of waste foundry sand stabilized with alkali-activated sugar cane bagasse ash-eggshell lime binder. *Construction and Building Materials*, *383*. https://doi.org/10.1016/j.conbuildmat.2023.131313.

Ferrazzo, S. T., Tonini de Araújo, M. & Consoli, N. C. (2024). Which solution is more sustainable: Waste foundry sand stabilized with alkali-activated binder or Portland cement? *Journal of Building Engineering, 84,* 108448. https://doi.org/10.1016/j.jobe.2024.108448.

Guney, Y., Aydilek, A. H. & Demirkan, M. M. (2006). Geoenvironmental behavior of foundry sand amended mixtures for highway subbases. *Waste Management*, *26*(9), 932–945. https://doi.org/10.1016/j.wasman.2005.06.007.

Guney, Y., Sari, Y. D., Yalcin, M., Tuncan, A. & Donmez, S. (2010). Re-usage of waste foundry sand in high-strength concrete. *Waste Management*, *30*(8–9), 1705–1713. https://doi.org/10.1016/j.wasman.2010.02.018.

Iqbal, M. F., Liu, Q. F. & Azim, I. (2019). Experimental study on the utilization of waste foundry sand as embankment and structural fill. *IOP Conference Series: Materials Science and Engineering*, *474*(1). https://doi.org/10.1088/1757-899X/474/1/012042.

Iqbal, M. F., Liu, Q. Feng, Azim, I., Zhu, X., Yang, J., Javed, M. F. & Rauf, M. (2020). Prediction of mechanical properties of green concrete incorporating waste foundry sand based on gene expression programming. *Journal of Hazardous Materials*, *384*, 121322. https://doi.org/10.1016/j.jhazmat.2019.121322.

Khatib, J. M., Herki, B. A. & Kenai, S. (2013). Capillarity of concrete incorporating waste foundry sand. *Construction and Building Materials*, *47*, 867–871. https://doi.org/10.1016/j.conbuildmat.2013.05.013.

Kumar, A. & Sharma, A. (2023). Improving strength characteristics of clayey soil incorporating waste foundry sand and recron 3S fiber. *AIP Conference Proceedings*, *2558*. https://doi.org/10.1063/5.0130703.

Lei n° 17.479, de 15 de janeiro de 2018. (2018). Dispõe sobre a utilização das areias descartadas de fundição (ADF). Assembléia Legislativa do Estado de Santa Catarina. https://www.legisweb.com.br/legislacao/?id=355620.

Lin, D.-F., Luo, H.-L., Lin, J.-D. & Zhuang, M.-L. (2018). Characterizations of temperature effects on sintered ceramics manufactured with waste foundry sand and clay. *Journal of Material Cycles and Waste Management, 20*, 127–136.

Mashifana, T. & Sithole, T. (2020). Recovery of Silicon Dioxide from Waste Foundry Sand and Alkaline Activation of Desilicated Foundry Sand. *Journal of Sustainable Metallurgy*, *6*(4), 700–714. https://doi.org/10.1007/s40831-020-00303-5.

Mastella, M. A., Gislon, E. S., Pelisser, F., Ricken, C., Silva, L. da, Angioletto, E. & Montedo, O. R. K. (2014). Mechanical and toxicological evaluation of concrete artifacts containing waste foundry sand. *Waste Management*, *34*(8), 1495–1500. https://doi.org/10.1016/j.wasman.2014.02.001.

Matos, P. R. de, Marcon, M. F., Schankoski, R. A. & Prudêncio, L. R. (2019). Novel applications of waste foundry sand in conventional and dry-mix concretes. *Journal of Environmental Management*, *244*, 294–303. https://doi.org/10.1016/j.jenvman.2019.04.048.

Matos, P. R., Pilar, R., Bromerchenkel, L. H., Schankoski, R. A., Gleize, P. J. P. & de Brito, J. (2020). Self-compacting mortars produced with fine fraction of calcined waste foundry sand (WFS) as alternative filler: Fresh-state, hydration and hardened-state properties. *Journal of Cleaner Production*, *252*, 119871. https://doi.org/10.1016/j.jclepro.2019.119871.

Morais, M., Levandoski, W. M. K., Reis, J. B., Rosa, F. D. & Korf, E. P. (2023). Environmental and technical feasibility of a waste foundry sand applied to pavement granular layers. *Soils and Rocks*, 46(1), 1–10. https://doi.org/10.28927/SR.2023.001722.

Navarro-Blasco, Í., Fernández, J. M., Duran, A., Sirera, R. & Álvarez, J. I. (2013). A novel use of calcium aluminate cements for recycling waste foundry sand (WFS).

Construction and Building Materials, *48*, 218–228. https://doi.org/10.1016/j.conbuildmat.2013.06.071.

Osinubi, K. & Moses, G. (2015). The influence of compactive effort on the desiccation-induced volumetric shrinkage of compacted bagasse ash treated foundry sand as hydraulic barrier material. *Jordan Journal of Civil Engineering*, *9*(4), 411–420.

Patel, M. D. & Chauhan, A. R. (2020). Stabilization of Soil by Foundry Sand Waste. *International Research Journal of Engineering and Technology*, *May*, 5675–5680. www.irjet.net.

Paul, P., Apedo, K. L., Belhaj, E. & Feugeas, F. (2022). Effect of waste chemical foundry sand on mechanical properties and damage process of cementitious materials using digital image correlation technology. *Journal of Cleaner Production*, *368*, 133103. https://doi.org/10.1016/j.jclepro.2022.133103.

Pelisser, G., Ferrazzo, S. T., Mota, J. D., dos Santos, C. P., Pelisser, C., Rosa, F. D. & Korf, E. P. (2023). Rice husk ash-carbide lime as an alternative binder for waste foundry sand stabilization. *Environmental Science and Pollution Research*, *30*(14), 42176–42191. https://doi.org/10.1007/s11356-022-25117-8.

Penkaitis, G. & Sígolo, J. B. (2012). Waste foundry sand. Environmental implication and characterization. *Geologia USP - Serie Cientifica*, *12*(3), 57–70. https://doi.org/10.5327/Z1519-874X2012000300004.

Philip, S., Ajin, A., Shahul, F., Babu, L. & Tina, J. (2019). Fabrication and performance evaluation of lightweight eco-friendly construction bricks made with fly ash and bentonite. *International Journal of Engineering and Advanced Technology*, *8*(6), 2090–2096. https://doi.org/10.35940/ijeat.F8478.088619.

CONSEMA. (2013) Resolução CONSEMA nº 26, de 06 de setembro de 2013. Estabelece as diretrizes sobre a utilização das Areias Descartadas de Fundição - ADF e adota outras providências. Conselho Estadual do Meio Ambiente, Governo do Estado de Santa Catarina. https://www.legisweb.com.br/legislacao/?id=258568#:~:text=Estabelece%20as%20diretrizes%20sobre%20a,incisos%20III%20e%20VII%2C%20art..

Sabour, M. R., Derhamjani, G. & Akbari, M. (2022). Mechanical, durability properties, and environmental assessment of geopolymer mortars containing waste foundry sand. *Environmental Science and Pollution Research*, *29*(16), 24322–24333. https://doi.org/10.1007/s11356-021-17692-z.

Sabour, M. R., Derhamjani, G., Akbari, M. & Hatami, A. M. (2021). Global trends and status in waste foundry sand management research during the years 1971-2020: A systematic analysis. *Environmental Science and Pollution Research*, *28*(28), 37312–37321. https://doi.org/10.1007/s11356-021-13251-8.

Şahmaran, M., Lachemi, M., Erdem, T. K. & Yücel, H. E. (2011). Use of spent foundry sand and fly ash for the development of green self-consolidating concrete. *Materials and Structures*, *44*(7), 1193–1204. https://doi.org/10.1617/s11527-010-9692-7.

Sgarlata, C., Ariza-Tarazona, M. C., Paradisi, E., Siligardi, C. & Lancellotti, I. (2023). Use of Foundry sands in the production of ceramic and geopolymers for sustainable construction Materials. *Applied Sciences*, *13*(8). https://doi.org/10.3390/app13085166.

Siddique, R. & Noumowe, A. (2008). Utilization of spent foundry sand in controlled low-strength materials and concrete. *Resources, Conservation and Recycling*, *53*(1–2), 27–35. https://doi.org/10.1016/j.resconrec.2008.09.007.

Siddique, R., Schutter, G. de & Noumowe, A. (2009). Effect of used-foundry sand on the mechanical properties of concrete. *Construction and Building Materials*, *23*(2), 976–980. https://doi.org/10.1016/j.conbuildmat.2008.05.005.

Siddique, R. & Singh, G. (2011). Utilization of waste foundry sand (WFS) in concrete manufacturing. *Resources, Conservation and Recycling*, *55*(11), 885–892. https://doi.org/10.1016/j.resconrec.2011.05.001.

Siddique, R., Singh, G. & Singh, M. (2018). Recycle option for metallurgical by-product (Spent Foundry Sand) in green concrete for sustainable construction. *Journal of Cleaner Production*, *172*, 1111–1120. https://doi.org/10.1016/j.jclepro.2017.10.255.

Siddiquea, R., Kaur, G. & Rajor, A. (2010). Waste foundry sand and its leachate characteristics. *Resources, Conservation and Recycling*, *54*(12), 1027–1036. https://doi.org/10.1016/j.resconrec.2010.04.006.

Singh, G. & Siddique, R. (2012). Effect of waste foundry sand (WFS) as partial replacement of sand on the strength, ultrasonic pulse velocity and permeability of concrete. *Construction and Building Materials*, *26*(1), 416–422. https://doi.org/10.1016/j.conbuildmat.2011.06.041.

Sithole, N. T. & Nkosi, M. T. M. (2022). Effect of Synthesis Conditions on Mechanical Strength of Aluminium Slag Modified Waste Foundry Sand Geopolymers. *IOP Conference Series: Earth and Environmental Science*, *1009*(1). https://doi.org/10.1088/1755-1315/1009/1/012011.

Sithole, N. T., Tsotetsi, N. T., Mashifana, T. & Sillanpää, M. (2022). Alternative cleaner production of sustainable concrete from waste foundry sand and slag. *Journal of Cleaner Production*, *336*. https://doi.org/10.1016/j.jclepro.2022.130399.

STATISTA. (2020). *Global casting production worldwide, 2018 to 2020*. https://www.statista.com/statistics/237526/casting-production-worldwide-by-country/.

Thaarrini, J. & Ramasamy, V. (2016). Properties of foundry sand, ground granulated blast furnace slag and bottom ash based geopolymers under ambient conditions. *Periodica Polytechnica Civil Engineering*, *60*(2), 159–168. https://doi.org/10.3311/PPci.8014.

Tittarelli, F. (2018). Waste foundry sand. In *Elsevier eBooks* (pp. 121–147). Elsevier Ltd. https://doi.org/10.1016/B978-0-08-102156-9.00004-3.

Torres, A., Bartlett, L. & Pilgrim, C. (2017). Effect of foundry waste on the mechanical properties of Portland Cement Concrete. *Construction and Building Materials*, *135*, 674–681. https://doi.org/10.1016/j.conbuildmat.2017.01.028.

United States Environmental Protection Agency. (2006). *State Toolkit for Developing Beneficial Reuse Programs for Foundry Sand*. United States Environmental Protection Agency. https://archive.epa.gov/sectors/web/pdf/toolkit.pdf.

Vinoth, M., Sinha, A. K., Guruvittal, U. K. & Havanagi, V. G. (2022). Strength of Stabilised Waste Foundry Sand Material. *Indian Geotechnical Journal*, *52*(3), 707–719. https://doi.org/10.1007/s40098-021-00586-9.

Yaghoubi, E., Arulrajah, A., Yaghoubi, M. & Horpibulsuk, S. (2020). Shear strength properties and stress–strain behavior of waste foundry sand. *Construction and Building Materials*, *249*, 118761. https://doi.org/10.1016/j.conbuildmat.2020.118761.

Yazoghli-Marzouk, O., Vulcano-Greullet, N., Cantegrit, L., Friteyre, L. & Jullien, A. (2014). Recycling foundry sand in road construction-field assessment. *Construction and Building Materials*, *61*, 69–78. https://doi.org/10.1016/j.conbuildmat.2014.02.055.

Zahmak, A., Abdallah, M., Jarah, B. & Arab, M. G. (2021). Environmental performance of alkali-activated binders for ground improvement. *Transportation Geotechnics*, *31*, 100631. https://doi.org/10.1016/j.trgeo.2021.100631.

Zhang, Y., Sappinen, T., Vilenius, M. & Korkiala-Tanttu, L. (2021). Investigations into stabilized structures with the use of waste foundry sand. *IOP Conference Series: Earth and Environmental Science*, *710*(1). https://doi.org/10.1088/1755-1315/710/1/012028.

Chapter 7

Development of Alkali-Activated Binders from Red Ceramic Wastes

Andres Lotero[*]
Cindy Johanna Moncaleano
and Nilo Cesar Consoli
Department of Civil Engineering, Universidade Federal do Rio Grande do Sul, Porto Alegre, RS, Brazil

Abstract

Red ceramic wastes (RCW) have been catalogued by several studies as a potential precursor material in the production of alkali-activated cements (AAC), thanks to their composition (aluminosilicate content) and semi-crystalline structure, whose reactive amorphous contents are due to the thermal history from the initial production process of red ceramics. The pozzolanic activity of these wastes is due to the destruction of the crystalline networks of the clay minerals (phyllosilicates) through a dehydroxylation process that occurs at temperatures between 600-900°C. Although this has been extensively demonstrated, thermal curing at temperatures between 60 and 90°C has been required to obtain representative mechanical responses of these red ceramic waste-based alkali-activated cements (RCWAAC) in relatively short curing periods. Additionally, they have shown the need for a higher alkali concentration and pH levels for their activation process, and in some cases, the inclusion of soluble silica (from alkaline silicate solutions or "waterglass" to the system) being these the conditions for the alkaline activation process of aluminosilicate precursors with low or no calcium

[*] Corresponding Author's Email: andreslotero@hotmail.com.

In: Environmental Geotechnics
Editors: E. Pavan Korf, P. Domingos Marques Prietto, C. Müller et al.
ISBN: 979-8-89113-651-9

content. On the other hand, different studies have reported the positive effect of the inclusion of additives of cementitious nature and/or calcium sources in RCWAAC, allowing the production of hybrid, binary, ternary or blended cementitious systems with very high mechanical performance, in shorter curing times, especially when cured at room temperature. This book chapter compiles the current state of the art regarding the production of RCWAAC from various types of alkaline activators and the inclusion or blending with other precursors to enhance their mechanical performance under different curing conditions.

Keywords: red ceramic wastes, alkaline activation, alkali-activated cement, hybrid alkaline cement

Introduction

In recent decades, there has been a great social and environmental concern about waste management in general. This is a consequence of the remarkable industrial and economic growth of societies. Particularly, construction and demolition waste (CDW) has reached quite worrying levels of generation, establishing its management and final disposal as a priority (Larrosa et al. 2023; Lotero et al. 2021; Komnitsas, 2011). Actions to address this problem have been focused on generating policies that promote the recovery of resources through reuse, recycling and the development of systems for converting waste into energy. The purpose of this is to convert waste into new resources or raw materials, which would initially be wasted (i.e., their insertion into the production chain), and thus reduce their social and environmental impacts (Gartner, 2004).

Ceramic products are produced from natural inorganic materials containing a high proportion of clay minerals after a process of dehydration and controlled firing at high temperatures (above 600ºC). Ceramic wastes are generated both by the ceramic production industry and by the construction and demolition sector, which defines a wide range of origin and composition, distinguishing different sources of these wastes, such as: structural ceramic products (bricks, tiles, vaults, etc.), ceramic tile, tiles, stoneware ceramics (cladding ceramics), among others (Barreto et al. 2023; Juan-Valdes et al. 2010; Puertas, 2008). Other authors classify them according to the type of clays used in the production process, red ceramic (from red clays) and white ceramic (from white clays) (Puertas et al. 2006; Pacheco-Torgal & Jalali, 2010), being mainly the Fe_2O_3 content that defines this color. Once their useful life is over, most of these ceramic materials are destined to be treated as waste, through final disposal in landfills. Statistics show that in some countries more than 95% of the total ceramic waste generated has this final destination (Cherene et al. 2023).

Red ceramic is a material that comes from the controlled calcination of red clay [$Fe_2O_3 > 2$ wt% (Xiao et al. 2012)] for the manufacture of elements such as bricks, blocks, tiles, among others, and is widely used in construction. The percentage of RCW present in CDW ranges on average between 45 and 54 wt%, however, it can reach values of more than 70% in developing countries (Silva et al. 2009). The red ceramic products manufacturing industry is also considered an important contributor to the generation of these wastes. Sintered products with deformations, breakage or over-firing are considered solid waste from the production process. The European Union ceramics industry reports that approximately one third of the total production of red ceramic pieces are disposed of without any treatment, i.e., they are deposited in landfills (Robayo, et al. 2016). On the other hand, China, India and Brazil, considered as the main ceramic producing countries in the world, report waste of up to 30% of the total volume of daily ceramic pieces manufactured (Samadi et al. 2020).

Given the high aluminosilicate content of red ceramic wastes (RCWs) and their semi-crystalline structure, several studies have reported their potential pozzolanic reactivity (Daigo et al. 2018). This is due to the dehydroxylation process of the clay minerals during sintering. However, low controls during the manufacture of ceramic pieces, excess and high firing temperatures tend to decrease the pozzolanic reactivity of RCWs (Medina et al. 2012). Therefore, RCWs have been categorized as a low reactivity pozzolanic material (Gaibor et al. 2023; Huseien et al. 2019). In this sense, the use of RCW as a pozzolan and supplementary material in the production of conventional cements has been quite limited (Mbereyaho et al. 2019). In addition, its high porosity, low density and mechanical strength have also questioned its reuse as an aggregate for the construction of structural backfill layers and the production of concrete (Magbool, 2022; Ray etal., 2021).

In this context, alkali-activated materials have been considered as a new technology for the alternative production of cementitious materials. In response to the environmental and social problems that conventional cements (e.g., Portland cement) have evidenced, mainly in the production process, and whose projected impacts tend to intensify due to their increasing mass use (Duxson et al. 2007a, b; Krivenko, 2017; Pacheco-Torgal et al. 2008). Alkaline or alkali-activated cements (AACs) have been characterized by their low CO_2 emissions and energy consumption during production, in addition to the possibility of including waste and by-products from various industries as precursors and activators for their manufacture (see Chapter 4). This makes it possible to reduce the exploitation of natural resources to obtain raw materials and promote the circular economy, with mechanical performance comparable to conventional cements (Provis, 2018; Van Deventer et al. 2010). The above has recently aroused great interest in the scientific community, encouraging the production of research works in the search for the feasibility of using AACs as additives for the chemical and/or chemical-mechanical stabilization of various geomaterials, and at the same time, the use of different geomaterials to be included as raw materials in their manufacture.

RCWs can be considered as precursor materials for the synthesis of alkaline cements (after being finely ground) thanks to their potential pozzolanic activity and the strong alkaline conditions in which the reactions take place (Amin et al. 2017; Gavali et al, 2019; Robayo et al. 2016). For the production of RCW-based alkaline cements (RCWAAC), it is necessary to generate systems (precursor-activator mixtures) of high pH levels (usually, pH > 13) by incorporating concentrated alkaline salts for RCW activation. In addition, in some cases, the inclusion of soluble silica from the activator and thermal curing to accelerate the reactions (Fořt et al. 2020). This is to counteract the low pozzolanic reactivity characteristic of RCWs and to achieve representative mechanical performance in the short term. Therefore, RCWAACs meet the conditions characteristic of alkaline activated precursors in a low-calcium model or systems (Rakhimova et al. 2021; Reig et al. 2017).

On the other hand, the inclusion of additives and/or supplementary materials, of a cementitious or pozzolanic nature, to form RCW-based admixture and hybrid cementitious systems, have reported a positive effect on RCWAACs. Improvements in mechanical performance have been recorded under different curing conditions, especially at ambient temperatures. This has allowed enhancing its practical applicability for in-situ (Lotero et al. 2023; Rakhimova, 2022) works. In this chapter, the advances in the use of RCW or its combination with other precursors (traditional and alternative) are summarized, emphasizing the effects of the nature and concentration of the alkaline activator, the effects of the curing conditions and the characteristics of the resulting reaction products.

Alkaline Activation of Crude RCW

Overview of the Mechanism of CAA Synthesis

AACs are synthesized by a condensation polymerization (polycondensation) chemical reaction, generically called "geopolymerization" (Davidovits, 2008), where a finely ground solid material rich in aluminosilicates in amorphous phase (reactive material soluble in an alkaline medium), called precursor, is mixed with an alkaline activator (usually alkali metal compounds) in order to substantially increase the pH of the mixture and start the different stages of the reaction. Therefore, two components are essential for the production of AACs, a solid precursor, usually referred to as the cementitious component, and a highly soluble alkaline activator, either in liquid or solid state (Shi et al. 2003, 2011). The alkaline activator serves the function of accelerating the solubility of the solid precursor material. It also favors the formation of stable hydrates (low solubility) and helps the formation of a compact structure with the hydrates formed (Garcia-Lodeiro et al. 2015). These alkaline activators can be mixed with the precursors in liquid state through alkaline solutions at a certain concentration (Two-part method), or in solid state (One-part method) to later add water, in the same way as conventional cements are dosed (Luukkonen et al. 2018).

In general, the mechanism of the chemical reaction of geopolymerization was divided into three stages by Glukhovsky (Glukhovsky, 1967), considered as the first model to describe the reaction mechanisms: (1) Destruction-Coagulation, consisting of the dissociation of the solid precursor upon contact with the alkaline activator, by breaking the ionic and covalent bonds present in the glassy phase of the precursor, where various alumino/silicate species are dissolved and pass in ionic state into solution, creating a supersaturated solution of aluminosilicates; (2) Coagulation-Condensation, at this stage the dissolved ionic species form a coagulation structure (gel) in the high alkalinity medium, favoring the condensation reaction and thus the formation of increasingly larger oligomeric species; (3) Condensation-Crystallization, corresponds to the polycondensation reactions, allowing the formation of the first microparticles of the primary reaction products (cementitious gels), which through continuous reorganization increase the connectivity of the gel network, leading to the precipitation of a high structural order aluminosilicate network (sometimes three-dimensional) that subsequently hardens (Van Deventer et al. 2007). However, different authors have extended this

Glukhovsky model by applying recent knowledge on zeolite synthesis to explain the formation of these inorganic polymers in more detail (Duxson et al, 2007a; Pacheco-Torgal et al. 2008).

This reaction mechanism, as well as its yield and the composition of the final reaction products, can vary depending on the composition and microstructure of the precursor material, the nature and concentration of the alkaline activator and the curing conditions. This has led to the development of a wide variety of AACs (Palomo et al. 2014) (see Chapter 4). However, some authors (Bakharev et al. 2000; Fernández-Jiménez, 2000; Garcia-Lodeiro et al. 2015; Provis & Van Deventer, 2009) state that AACs can be grouped mainly into three categories or models of alkaline activation according to the nature of their cementitious component (CaO-SiO_2-Al_2O_3 system). The first corresponds to the system or model of alkaline activation of high-calcium materials (system $[Na,K]_2O$-CaO-Al_2O_3-SiO_2-H_2O), characterized by including precursor materials rich in calcium and silicon in amorphous phase ($SiO_2 + CaO > 70\%$) (Wang et al. 1995). The main synthesis product corresponds to a calcium silicate hydrate gel very similar to the cementitious gel produced by the hydration of Portland clinker, which absorbs aluminum in its structure (C-A-S-H gel) and at the same time presents a higher structural order or degree of polymerization (Puertas et al. 2011).

The second corresponds to the alkaline activation model of low calcium materials (system $[Na,K]_2O$-Al_2O_3-SiO_2-H_2O), characterized by including precursor materials rich in silicon and aluminum ($CaO < 10\%$) (Komnitsas et al. 2007). In this model, more aggressive working conditions are required for the initiation of the reactions, i.e., highly alkaline media and curing conditions at temperatures that can vary between 60° and 200°C (Fernández-Jiménez et al. 2006; McLellan et al. 2011). The main reaction product is an alkaline aluminosilicate hydrate gel [usually an N-A-S-H or K-A-S-H gel, depending on the nature of the activator (Fernández-Jiménez et al. 2006)] with cementitious properties, corresponding to a three-dimensional inorganic alkaline polymer of high structural order, also called "geopolymer" (McLellan et al. 2011; Van Deventer et al. 2007).

Finally, the third alkaline activation model corresponds to a combination of the previous ones, commonly called blended or hybrid alkaline cement, as a result of the alkaline activation of precursors with CaO, SiO_2 and Al_2O_3 contents $> 20\%$, which may include mixtures of different pozzolans and/or the addition of conventional cements as precursors (García-Lodeiro et al. 2016). The primary reaction products are a mixture of cementitious gels that coexist in the same cementitious matrix, including a C-A-S-H gel that can absorb sodium ions and an (N,C)-A-S-H gel corresponding to an N-A-S-H gel with high calcium content, which can polymerize or depolymerize with time and pH of the system (Donatello et al. 2014; Palomo et al. 2013).

Reactivity of RCWs for RCWAAC Production

RCWs basically derive from the production process of ceramic materials and from construction and demolition activities. The waste from the ceramic production industry, before firing, is generally due to breakage and deformation and is characterized by the fact that the intrinsic properties of the raw material are not affected. Usually the defective pieces before the sintering stage are subjected to a segregation process to be later reinserted into the production line (Bravo et al. 2020). On the other hand, waste from firing defects, due to excessive heat (over-firing) or insufficient heat (under-firing), is characterized by modifying the physicochemical properties of the ceramic product. Therefore, sintered products with firing defects are considered as solid residues of the industrial process (Gautam et al. 2021).

Regarding the waste generated by the construction sector, the red ceramic fraction represents the highest percentage of all the materials present in the CDW. According to the average compositional distribution of CDW, the stone fraction corresponds to 75 wt% of the CDW, where the RCWs on average reach 54 wt% (Koyuncu et al. 2004). Pozzolans are defined as materials (inorganic, natural or artificial) containing silica and/or aluminosilicates in reactive phases, which, although they do not exhibit hydraulic behavior by themselves, when finely divided and in the presence of moisture, react chemically with calcium hydroxide ($Ca(OH)_2$). These reactions occur at room temperature and produce a new material with cementitious properties (analogous to that generated in the hydration of Portland cement), in a process known as pozzolanic reaction (Dodson, 1990).

The term 'pozzolanic activity' includes two parameters, the maximum amount of lime that can react with a pozzolan (combined or fixed lime) and the rate at which such reactions occur. Both factors depend on the nature of the pozzolans and, more precisely, on the quality and quantity of the vitreous or reactive phases (Weise et al. 2023). Heat-treated clays can be considered as one of the first materials used as pozzolana (Hong, et al. 2023). Since the Phoenician civilization, there are reports of the use of pulverized clay bricks to obtain hydraulic mortars. Subsequently, Roman constructions, through their remains all over Europe, testify how from the mixture of brick powder with hydrated lime they obtained materials with binding properties (Salustio, et al. 2022).

According to Huertas, et al. (1997) and Heller-Kallai (2013), clay minerals when subjected to a thermal process, undergo a series of transformations which begin with their dehydration, the appearance of a vitreous phase, and finally, the formation of new crystalline phases. All these alterations have a strictly evolutionary character, of decomposition (the first two) and formation (the last one), which brings about the appearance of a new material whose structure could be defined as that of a polycrystalline solid, with a potentially amorphous and glassy matrix (Wachtman, et al. 2009). This evolutionary process depends largely on the mineralogical composition of the clay and the heat treatment conditions. Table 1 summarizes the behavior of different types of clay minerals during calcination.

The loss of free and adsorbed water is known as the clay dehydration process (Heller-Kallai, 2006). On the other hand, the dehydroxylation process is a collapse of the clay mineral structure, causing a distortion and, therefore, the formation of an amorphous structure (Drits et al. 2012). For example, in the case of kaolinite, its dehydroxylation leads to the formation of metakaolin (at temperatures between 500 to 900°C). Metakaolin is characterized by a predominantly amorphous structure (Sperinck et al. 2011). The dehydroxylation process of the clay is a relevant aspect in its pozzolanic activity and depends mainly on the type or group of clay mineral (Zhu et al. 2022). Under these conditions of disordered lattice structure, silica and alumina present in the clay can bind calcium hydrate. However, heat treatments at temperatures above 900°C, these aluminosilicates can reorganize and form thermodynamically more stable compounds (e.g., mullite, tridymite, diopside, wollastonite, gehlenite, plagioclase, among others). These formed minerals are known as high-temperature phases and do not exhibit pozzolanic activity (Sudo et al. 1967).

By means of pozzolanicity analysis, Baronio & Bindat (1997) evaluated finely ground ceramic pieces from the ancient basilica of San Lorenzo Maggiore located in Milan and built in the 4th century (Byzantine Empire). The results were contrasted with RCWs collected from a factory in Milan, considered as ceramic material coming from a modern structure and which were fired at temperatures above 900°C. The results showed that the medieval RCWs developed pozzolanic activity. However, RCWs manufactured at firing temperatures above 900°C did not show pozzolanic activity. The study concludes that not all finely ground RCWs can be considered pozzolanic.

Table 1. Behavior of some clay minerals during heat treatment. Adapted from Heller-Kallai (2013)

Process	Kaolinite	Illite	Smectite	Chlorite
H_2O loss adsorbed	-	100°-200°C	110°-200°C	-
Dehydroxylation	450°-525°C	600°-700°C	500°-800°C	500°-800°C
Destruction of crystalline network	550°C	850°C	850°C	850°C
Spinel's formation	950°C	900°C	1100°C	1200°C
Olivino's formation	-	-	-	800°-900°C
Mullita's formation	1000°C	1100°C	1000°C	-
Enstatite's formation	-	-	1000°C	1200°C
Cristobalite's formation	1300°C	-	900°C	-

De Rojas et al. (2001), based on Fratini (BS EN 196-5, 2011), compressive strength and Langavant calorimetry (BS EN 196-9, 2010) tests, evaluated the pozzolanic activity of finely ground RCW (3,500-4,000 Blaine units) and compared them with two high activity pozzolans (silica fume and fly ash). The study reports low pozzolanic activity of RCWs at early ages (<28 days). At long curing times, the RCWs evidenced pozzolanic activity in magnitudes comparable to the reference pozzolans. However, the heat of hydration of the pozzolanic reaction of the RCWs was lower. Puertas et al. (2006) studied 6 different types of RCW from different industrial ceramic plants in Europe. Chemical characterization tests established that RCWs are neither toxic nor radioactive. Pozzolanicity analyses verified the onset of pozzolanic activity of RCWs (in $Ca(OH)_2$ saturated solutions) after the first 8 days. This was due to the presence of silicates (SiO_2) and aluminates (Al_2O_3) in glassy phases. XRD analysis allowed quantifying variable contents of these aluminosilicates, between 5 and 36 wt%. The study concludes on the potential use of RCW as a precursor material in the production of RCWAAC and the need to implement thermal curing to improve the kinetics of the reactions.

Several studies (De Rojas et al. 2001, 2006; Frías et al. 2008) have reported that in order to use RCW as a pozzolanic additive, it must be ground to a specific surface area, or Blaine value, of at least 3500 cm^2/g. In addition, they conclude that RCW, although showing pozzolanic activity, after one day of curing, the percentage of lime fixed is very low (approximately 19% of the total lime available) compared to the values recorded for other pozzolans such as silica fume and fly ash. However, after 90 days, very similar results are obtained to those of the highly reactive reference pozzolans. On the other hand, Juan-Valdes et al. (2010) show that RCWs have a chemical composition similar to other pozzolanic materials, such as silica fume, fly ash and some metallurgical slags. Silica, aluminum oxide and iron oxide predominate (approx. 75.5%). In addition to CaO contents of 12.41%, alkalis of 4.22%, sulfates (SO_3) of 0.79% and small ignition losses in the order of 3.44%. On the other hand, the mineralogical composition, from X-ray diffraction (XRD), reported the presence of quartz, muscovite, calcite, microcline and anorthite phases. The study concludes that the firing temperatures used for the production of ceramic material (usually around 900ºC) were sufficient and suitable to activate the clay minerals.

Some studies (Jain et al. 2022; Jiménez et al. 2013; Katzer, 2013; Magbool, 2022; Medina et al. 2012; Puertas, 2008), agree that partial replacements of cement by RCW in the production of concretes and mortars, have exhibited positive responses. However, workability, compressive, tensile and flexural strength decrease as the replacement ratio increases (replacements <10-20 wt%, do not alter the properties of the cementitious material). However, concretes with RCW additions, evidence higher durability, as a product of pozzolanic reactions of RCWs with Portlandite formed by the hydration of Portland cement (Brekailo et al. 2022).

Keppert et al. (2018) evaluated the reactivity of two RCW samples from XRD and XRF assays. Reporting percentages of aluminosilicates in amorphous phase of 27.8% and 47%. The authors conclude that, even with a glass content of less than 30%, RCWs can develop strengths of up to 38 MPa after 28 days of curing at room temperature (23°C) by alkaline activation with waterglass (mixture of NaOH and Na_2SiO_3). This was achieved by optimizing the mixtures according to the SiO_2/Al_2O_3 and $Al_2O_3/(Na_2O + K_2O)$ ratios, with values of 2.8 and 1.0, respectively, calculated based on the composition of the alkaline activating solution and the amorphous portion present in the RCWs.

Ruiz-Santaquiteria et al. (2013) by means of a test procedure raises the possibility of quantifying the total fraction of aluminosilicates present in reactive phases (i.e., Al_2O_3 and SiO2 in amorphous phase) of pozzolans with low calcium contents. The procedure consists of subjecting a pozzolan sample (1.0 g) to an acid attack with a 100 ml solution of hydrofluoric acid at 1% by volume, keeping the mixture in constant agitation for 5 hours at room temperature. Subsequently, by means of a vacuum filtration with cellulose acetate filter paper (of 2 μm pore size), the insoluble solid residue retained on the filter paper was separated from the leached solution containing the silica and alumina in reactive phases of the pozzolan.

Lotero et al. (2023), following the procedure described by Ruiz-Santaquiteria et al. (2013), determined by atomic absorption tests that approximately 18.7 wt% of silicates and 8.6 wt% of aluminates present in RCWs are in the glassy or reactive state. An average reactive SiO_2/Al_2O_3 ratio of 2.12 wt% was reported. The authors concluded on the suitability of RCWs (SiO_2/Al_2O_3 ratio > 2) in the formation of cementitious reaction products of a high degree of polymerization after the alkaline activation process. However, the low content of total reactive aluminosilicate phases (approximately 27 wt%) present in the RCWs potentially determines the formation of a low amount of these reaction products, characterizing the RCWs as low reactivity precursors for the production of AAC.

Factors Influencing RCWAACs

Synthesis of AAC, in general, is affected by a wide variety of factors, however, the composition and fineness of the solid precursor, the nature and concentration of the alkaline activator, the amount of free water in the system (water/binder ratio) and the curing conditions can be considered as the most relevant (Lotero et al. 2020; Servi et al. 2022).

Different studies have reported that initial thermal cures at temperatures between 60 and 90°C are required to obtain representative mechanical responses of RCWAACs in relatively short curing periods (7 and 28 days). The above, taking into account the low pozzolanic activity (preliminarily exposed) that RCWs experience at early ages, even, after the alkaline activation process with different salts of high solubility, such as, for example, NaOH and KOH (Lotero et al. 2023; Reig et al. 2013a, b; Robayo et al. 2016). The mechanical strengths of RCWAACs increase exponentially with curing temperature, decreasing the time needed to reach a target value, i.e., temperature is a catalyst for reactions, especially at early ages (first 48-72 hours post-synthesis) (Ferreira et al. 2022; Larrosa et al. 2023). Thermal curing accelerates the dissolution process of aluminosilicates and, additionally, allows a better development of the microstructure of the alkali cement formed (Mikuni, et al. 2007; Palomo et al. 2013, 2014).

Puertas et al. (2006) in the production of RCWAAC reported the need to implement thermal curing to improve the kinetics of the reactions for the formation of cementitious products. The RCWAAC pastes reached maximum strengths of up to 13 MPa after 8 days of curing in compression tests and 2.5 MPa in flexural tests, after being activated with sodium hydroxide (NaOH) and sodium silicate (Na_2SiO_3) solutions, independently, at 6 and 12 Molar concentrations. The study comments that, based on preliminary tests, it was necessary to perform a thermal curing during the first 24 hours at 85°C and a relative humidity of 99% in an oven, after the synthesis of the RCWAACs, in order to achieve considerable yields of the mixtures. After this time, the specimens were kept in a humid chamber until the age of testing (8 days) at a temperature of 21 ± 2°C and relative humidity of 99%. The study concluded that the amount of glassy phase and the nature of the activator did not have a strong influence on the development of the mechanical properties of the pastes. However, it reported the need for further investigations to understand the influence of other factors on RCWAAC production.

These findings have been confirmed, recently, by Silva et al. (2019) and Lotero et al. (2023), where satisfactory mechanical performances of RCWAAC have been reported after prolonged cures (up to 90 days) and/or the inclusion of thermal cures at temperatures up to 80-90°C, by using various alkaline activators. This is a characteristic of the system or model of alkaline activation in low calcium (Komnitsas et al. 2007). This has been a limitation for the use of RCWAAC in in-situ applications. However, it is considered a viable alternative in the manufacture of prefabricated elements, where the curing conditions can be controlled (Luhar et al. 2021).

Reig et al. (2013a, 2003b), proposed a study to understand the influence of the type and concentration of alkali activator on the strength and microstructure of finely ground RCW (mean particle size of 20.9 μm) alkali-activated with waterglass, after curing at 65°C for 3 and 7 days. The authors reported that the compressive strength and microstructure of RCWAACs depended on both the alkali (Na^+) concentration and the SiO_2/Na_2O ratio (or modulus of the activating solution - Ms). Compressive strength of up to 50 MPa after 7 days of curing was recorded for mixtures with high Na^+ concentrations. Furthermore, it was concluded that a low water/binder ratio (≈0.35) improved the performance of the mixtures.

Diverse authors (Fernandez Jimenez, 2000; Garcia-Lodeiro et al. 2015; Puertas, 1995) from the study of alkali-activated cements using alkaline hydroxides agree that both cations (Na^+) and anions (OH^-) of the alkaline solution are an essential part in the different phases of the co-polymerization process for the formation of these binders. The cations maintain the pH levels and form part of the gel structure (network modifiers), the anions induce the hydrolysis of the covalent bonds of the source alumino-silicates (dissolution phase) acting as catalysts for the reactions. However, highly concentrated solutions can generate phenomena similar to efflorescence due to the effects of a new crystallization of the alkaline hydroxides and carbonate phases (reaction with atmospheric CO_2) generating a negative effect on the hardening time as well as on the mechanical behavior.

The inclusion of soluble silica from the activator (e.g., sodium silicates) is a relevant factor that defines the kinetics (accelerates the reaction), structure and composition of the cementitious gel formed, mainly in its initial stages of formation. However, high concentrations of silica in the activating solution decrease the pH and increase its viscosity, which induces a decrease in the reaction rate of the solid precursor (Duxson et al. 2007a). Allahverdi & Kani (2009) from RCW activated with NaOH and Na_2SiO_3 solutions (Na_2O of 6, 7 and 8% and Ms of 0.6), fabricated RCWAAC and reported a compressive strengths up to 40 MPa (in the mixtures with 8% alkali contents) after 28 days of curing at room temperature (25°C). Tuyan et al. (2018) reported compressive strengths up to 36 MPa in RCWAAC with a waterglass solution of Ms equal to 1.6 and 10% Na_2O by weight of dry RCW powder after curing for 5 days at 90°C and 40% relative humidity.

On the other hand, excess water in the alkaline system (high water/binder ratios) favors the formation of capillary pores and, therefore, negative changes in the microstructure of the material (Provis et al. 2009; Torres-Carrasco, 2014). Unlike cementitious products formed by hydration of Portland cement, where water is chemically bound to the cementitious gel structure, water in AACs is found in the pores of the gel (Servi et al. 2022). Therefore, the fraction involved in the reactions is minimal. The hardening and evolution of cementitious products is directly related to the loss of water during curing (Yang et al. 2011).

Similarly, Sun et al. (2013) determined the thermal behavior (exposure to 100, 200, 400, 600, 800, and 1000°C) of RCWAACs synthesized using alkali hydroxides and/or sodium/potassium silicate solutions. The RCWAACs recorded maximum compressive strengths at 28 days of up to 71.1 MPa. Moreover, after heat treatment at 1000°C post-synthesis, the RCWAACs reached strengths above 75 MPa, evidencing high thermostability and heat resistance.

Robayo et al. (2016) established optimum compressive strength values using 10% Na_2O concentrations in the alkaline activation of RCW with NaOH solutions, reporting values of 7.49 MPa and 15.57 MPa in 28 days of curing by subjecting the mixtures to initial cures of 25°C and 70°C for 48 hours, respectively. On the other hand, additions of soluble silica in the activator (mixture: NaOH + Na_2SiO_3) significantly increased the strengths of the mixtures with 100% RCW, reporting a maximum strength of 54.38 MPa in 28 days cured at 25°C, which means an increase of 626% with respect to the strengths obtained in the pastes activated exclusively with NaOH.

Regarding the cementitious reaction products, taking into account the predominant aluminosilicate composition of RCWs and their low calcium content in reactive phases, several studies coincide in the formation of an alkaline aluminosilicate hydrate as the main reaction product, with an amorphous structure and a high degree of polymerization (three-dimensional structural network), compared to the cementitious gels formed by hydration of Portland cement (Abd Ellatief et al. 2023; Ricciotti et al. 2022). Where, the composition of the cementitious gel is mainly defined by the nature of the alkaline activator. RCW alkaline activated with hydroxides, carbonates and sodium silicates, form a hydrated sodium silico-aluminate sodium gel (N-A-S-H gel). RCW activated with hydroxides and potassium silicates tend to form a hydrated potassium silico-aluminate gel (K-A-S-H gel) (Hou et al. 2023; Mohana & Bharathi, 2023). Additionally, the formation of zeolites as secondary reaction products (Collins et al. 2020; Duxson et al. 2007a) has been reported.

RCW-Based Alkali-Activated Blended Cements

The literature on the production of AACs based on calcium-rich precursors has been mainly focused on granulated blast furnace slag (GBFS), with very limited work involving the use of other types of precursors. GBFS is an artificial geomaterial, considered as a by-product of the smelting of iron ores, coke and a flux (either limestone or dolomite) in the iron production process (Awoyera & Adesina, 2019). Like GBFS, research for the production of AACs from low calcium content materials has focused on the study of metakaolin (calcined kaolinite) and fly ash - F type (FA), establishing the principles of synthesis of aluminosilicate precursors based on these materials. This precursors (and their combinations) can be considered as the conventional sources of aluminosilicates for the production of AAC due to their high reactivity (Lotero et al. 2023; Wu et al. 2019). In addition, the vast majority of research has focused on the study of these precursors.

Despite the high performance shown by CAAs produced from GBFS, FA and metakaolin, they are currently highly required by the conventional cement producing industry for use as supplementary materials. Therefore, they present limitations in terms of availability, quantity and cost (Rivera et al. 2020). In this context, in recent years the need to search for new precursors has been addressed, especially sources of aluminosilicates from various wastes and/or low value-added by-products, becoming a recent line of research with the aim of globalizing and enhancing the competitiveness of AACs.

On the other hand, the use or addition of precursors of low or lower reactivity, such as RCWs, with conventional cements and/or precursors of high reactivity (blended cements or alkaline additives) has made RCWs viable and valuable for the production of alternative cements. In this sense, different researches have concluded on the positive effect of the inclusion of cementitious additives, high reactivity precursors (e.g., GBFS, FA and metakaolin) and/or calcium sources in RCWAACs, allowing the production of binary, ternary, admixture or blended cementitious systems with very high mechanical performance, in shorter curing times, especially when cured at room temperature (Hwang et al. 2019; Rakhimova et al. 2022).

In 2016, Reig et al. (2016) with the addition of calcium aluminate cement (40%) to RCWs, show that geopolymerization reactions are accelerated, obtaining strengths of 50 MPa at room temperature and 3 days of curing, after being alkaline activated with NaOH solutions. Robayo et al. (2016), in RCW mixtures with additions of ordinary Portland cement (20% by weight) and the inclusion of Na_2SiO_3 in the activating solution, achieved strengths of up to 102.6 MPa in 28 days of curing at room temperature. Zawrah et al. (2016) determined the effect of the substitution of granulated blast furnace slag (20, 40, 60, and 80%), characterized by having a composition rich in CaO (44.4% by weight) in the RCW alkaline activation process; the results show an evolution of the strength and decrease in the porosity of the pastes with the increase of the slag from 3 days of curing at room temperature.

Hwang et al. (2019) from alkali-activated pastes in mixtures of RCW, FA and GBFS, with silicate and sodium hydroxide solutions, have reported compressive strengths up to 70 MPa at room temperature in 28 days of curing. Similarly, Gaibor et al. (2020), from RCW pastes with FA and GBFS, activated with mixtures of sodium silicate and sodium hydroxide, report maximum compressive strengths of 20 MPa and 59 MPa, respectively, after 90 days of curing at room temperature. The addition of GBFS, allowed the alkali activator concentration to be much lower (2-8%). Increasing the addition of FA required a higher alkali concentration (5-20%).

Rakhimova & Rakhimov (2020) showed that the addition of 40% RCW improves the 7-day and 28-day strength of GBFS and GBFS-based pastes and mortar mixes after alkaline activation with sodium carbonate and sodium silicate. Furthermore, it shows that mixes with up to 60% slag replacements with RCW do not lose strength. Performance and structure of GBFS-based alkaline cement (GBFSAAC) is controlled primarily by the nature of the slag and the nature/dosage of the activator(s) used (Puertas et al. 2011). However, there is a consensus that the main reaction product is a C-A-S-H gel, with a disordered structure similar to the C-S-H gel, but with the absorption of aluminum (acting as a surrogate) to form a network in the form of cross-linked linear chains. Additionally, at the compositional level, this gel contains lower Ca/Si ratios, generally ranging from 0.9-1.2 (Wang & Scrivener, 1995).

Sarkar & Dana (2021), in alkali-activated mortars with waterglass, studied the progressive replacement of metakaolin by RCW (33.3%, 50% and 66.6%) on geopolymerization and its resulting properties. The findings showed that the lower replacement fraction (33.3%) did not reduce the compressive strength of the mixtures (>60 MPa). Additionally, they conclude that higher replacement fractions determine the formation of a lower amount of cementitious phase. Nevertheless, maximum compressive strength values of up to 30 MPa were achieved. This was attributed to the improved particle packing by the incorporation of RCWs. N-A-S-H/K-A-S-H cementitious gels, depending on the type of activator used, are the preponderant reaction products in alkali-activated systems based on metakaolin or FA (Duxson et al. 2007b).

Lotero et al. (2023) proposed the combination of two wastes, low reactivity RCW and carbide lime (CL), a by-product of acetylene gas production, to produce AAC from NaOH solutions. The authors conclude on the existence of a dosage of the residues (CL/(RCW+CL)) and alkali concentration (%Na_2O), which tends to maximize the mechanical properties of the pastes (CL/(RCW+CL)=0.3 and %Na_2O=4%), where higher proportions, the mechanical performance of the mixtures tends to decrease. Additionally, the authors report the formation of a C-A-S-H and (N,C)-A-S-H gel, which coprecipitate in the same cementitious matrix. These gels were characterized by developing a higher degree of polymerization (structural order), compared to the C-S-H/C-(A)-S-H gel formed by control pastes, as a result of a pozzolanic type reaction (RCW-CL mixtures activated with water).

Final Considerations

RCWs have been catalogued as a potential precursor material for the production of AAC. RCWs are characterized by their majority composition in aluminosilicates and by having a semi-crystalline structure, depending on their process and thermal history during the production of the ceramic pieces. That is, a fraction of their aluminosilicate content would potentially be in the vitreous or amorphous phase, available for alkaline activation. Several studies have reported reactive aluminosilicate contents in the order of 5 wt% to 36 wt%. Therefore, their high heterogeneity and the formation of new mineralogical phases during clay sintering have classified these wastes as having low pozzolanic reactivity, especially at early ages.

The alkaline activation of raw RCW through hydroxides and alkaline silicates to produce RCWAACs has been shown to be a viable alternative to enhance their reuse and valorization. It has been reported that RCWAACs achieve higher mechanical performance (especially in terms of durability) compared to RCW-based pozzolanic cements. This is a product of the polycondensation reactions of the dissolved aluminosilicates from the reactive phases of RCWs in a high alkalinity medium. The formation of cementitious gels with a high degree of polymerization (N-A-S-H/K-A-S-H) gel as the main product of the reaction and zeolites as secondary products was reported. However, factors such as the composition and particle size of the RCWs, the nature, concentration and modulus of the activating solution, as well as the water/binder ratio and the curing conditions determine the kinetics of the reactions and the final performance of the RCWAAC.

The inclusion of thermal curing (between 60 and 90°C) post-synthesizing (first 48 to 72 hours) of RCWAACs accelerates the dissolution phase of the amorphous phases (higher reaction kinetics) and improves the microstructure of the cementitious reaction products. Additionally, a low water/binder ratio not only decreases the initial hardening time of the mixtures, but also controls excess porosity of the hardened RCWAACs. Finally, the production of RCW-based blended or blended cements (e.g., GBFS, FA, metakaolin) have reported mechanical performance comparable to conventional cements when cured at room temperature, making their applicability for in-situ work feasible.

On the other hand, it is necessary to investigate the use of other types (or mixtures) of alkaline activators, especially the so-called green alkaline activators for the production of RCWAACs. That is, the use of binary or ternary mixtures of activators produced from wastes and/or by-products. This is because commercial activators have been considered corrosive, viscous, expensive and with a high energy and carbon component (Lotero et al. 2023; Núñez et al. 2023; Saldanha et al. 2023). Finally, studies focused on the life cycle analysis associated with the production and application of RCWAACs in civil construction engineering are suggested. The above, in order to evaluate the sustainability of these alternative cementitious materials.

References

Abd Ellatief, M., Abadel, A. A., Federowicz, K. & Abd Elrahman, M. (2023). Mechanical properties, high temperature resistance and microstructure of eco-friendly ultra-high performance geopolymer concrete: role of ceramic waste addition. *Construction and Building Materials, 401*, 132677. https://doi.org/10.1016/j.conbuildmat.2023.132677.

Allahverdi, A. & Kani, E. N. (2013). Use of construction and demolition waste (CDW) for alkali-activated or geopolymer cements. *Handbook of recycled concrete and demolition waste.* Woodhead Publishing Series in Civil and Structural Engineering. Chapter 18, pp. 439–475. https://doi.org/10.1533/9780857096906.3.439.

Amin, S. K., El–Sherbiny, S. A., El–Magd, A. A., Belal, A. & Abadir, M. F. (2017). Fabrication of geopolymer bricks using ceramic dust waste. *Construction and Building Materials, 157*, 610-620. https://doi.org/10.1016/j.conbuildmat.2017.09.052.

Bakharev T., Sanjayan J. G. & Cheng Y. B. (2000). Effect of admixtures on properties of alkali-activated slag concrete. *Cement and Concrete Research, 30*, 1367–1374. https://doi.org/10.1016/S0008-8846(00)00349-5.

Baronio, G. & Binda, L. (1997). Study of the pozzolanicity of some bricks and clays. *Construction and building Materials, 11(1)*, 41-46. https://doi.org/10.1016/S0950-0618(96)00032-3.

Barreto, G. N. S., Babisk, M. P., Delaqua, G. C. G., Gadioli, M. C. B., Monteiro, S. N. & Vieira, C. M. F. (2023). Study of the effect of the incorporation of two types of wastes and their combined effects into red ceramics. *Journal of Materials Research and Technology, 27*, 641-650. https://doi.org/10.1016/j.jmrt.2023.09.303.

Bravo, M., Duarte, A. P., De Brito, J., Evangelista, L. & Pedro, D. (2020). On the development of a technical specification for the use of fine recycled aggregates from construction and demolition waste in concrete production. *Materials, 13(19)*, 4228. https://doi.org/10.3390/ma13194228.

Brekailo, F., Pereira, E., Pereira, E., Farias, M. M. & Medeiros-Junior, R. A. (2022). Red ceramic and concrete waste as replacement of portland cement: Microstructure aspect of eco-mortar in external sulfate attack. *Cleaner Materials, 3*, 100034. https://doi.org/10.1016/j.clema.2021.100034.

BS EN - British Standards European Norm. (2010). *Methods of testing cement Heat of hydration. Semi-adiabatic method.* (BS EN 196-9:2010). European Committee for Standardization. https://www.cencenelec.eu.

BS EN - British Standards European Norm. (2011). *Methods of testing cement Pozzolanicity test for pozzolanic cement, European Standard specifies.* (BS EN 196-5:2011). European Committee for Standardization. https://www.cencenelec.eu.

Cherene, M. G. P., De Castro Xavier, G., Da Silva Barroso, L., Oliveira, J. D. S. M., De Azevedo, A. R. G., Vieira, C. M., Alexandre, J. & Monteiro, S. N. (2023). Technological and microstructural perspective of the use of ceramic waste in cement-based mortars. *Construction and Building Materials, 367*, 130256. https://doi.org/10.1016/j.conbuildmat.2022.130256.

Collins, F., Rozhkovskaya, A., Outram, J. G. & Millar, G. J. (2020). A critical review of waste resources, synthesis, and applications for Zeolite LTA. Microporous and mesoporous *Materials, 291*, 109667. https://doi.org/10.1016/j.micromeso.2019.109667.

Daigo, I., Kiyohara, S., Okada, T., Okamoto, D. & Goto, Y. (2018). Element-based optimization of waste ceramic materials and glasses recycling. *Resources, Conservation and Recycling, 133*, 375–384. https://doi.org/10.1016/j.resconrec.2017.11.012.

Davidovits, J. (2008). *Geopolymer Chemistry and Applications*, (Institut Géopolymère, 5th ed.) Saint-Quentin, France.

De Rojas, M. S., Frías, F., Rivera, J., Escorihuela, M. J. & Marin, F. P. (2001). Research about the pozzolanic activity of waste materials from calcined clay. *Materiales de construcción, 51(261)*, 45-52. https://doi.org/10.3989/mc.2001.v51.i261.379.

De Rojas, M. I., Marin, F., Rivera, J. & Frías, M. (2006). Morphology and properties in blended cements with ceramic wastes as a pozzolanic material. *Journal of the American Ceramic Society, 89(12)*, 3701-3705. https://doi.org/10.1111/j.1551-2916.2006.01279.x.

Dodson, V. H. (1990). Pozzolans and the pozzolanic reaction. In Concrete admixtures (pp. 159-201). Boston, MA: Springer US. https://doi.org/10.1007/978-1-4757-4843-7_7.

Donatello, S., Maltseva, O., Fernandez-Jimenez, A. & Palomo, A. (2014). The early age hydration reactions of a hybrid cement containing a very high content of coal bottom ash. *Journal of the American Ceramic Society*, 97, 929-937. https://doi.org/10.1111/jace.12751.

Drits, V. A., Derkowski, A. & McCarty, D. K. (2012). Kinetics of partial dehydroxylation in dioctahedral 2: 1 layer clay minerals. *American Mineralogist, 97(5-6)*, 930-950. https://doi.org/10.2138/am.2012.3971.

Duxson, P., Fernández-Jiménez, A., Provis, J. L., Lukey, G. C., Palomo, A. & Van Deventer, J. S. J. (2007a). Geopolymer technology: The current state of the art. *Journal of Materials Science, 42*, 2917–2933. https://doi.org/10.1007/s10853-006-0637-z.

Duxson, P., Mallicoat, S. W., Lukey, G. C., Kriven, W. M. & Van Deventer, J. S. J. (2007b). The effect of alkali and Si/Al ratio on the development of mechanical properties of metakaolin-based geopolymers. *Colloids and Surfaces A: Physicochemical and Engineering Aspects, 292 (1)*, 8-20. https://doi.org/10.1016/j.colsurfa.2006.05.044.

Ferreira, W. M., Cruz, A. S., de Azevedo, A. R., Marvila, M. T., Monteiro, S. N. & Vieira, C. M. F. (2022). Perspective of the application of ash from the ceramic industry in the development of alkali-activated roof tiles. *Ceramics International, 48(5)*, 6250-6257. https://doi.org/10.1016/j.ceramint.2021.11.166.

Fernández-Jiménez, A. (2000). *Cementos de escorias activadas alcalinamante: influencia de las variables y modelización del proceso*. Universidad Autónoma de Madrid.

Fernández-Jiménez, A., Palomo, A. & Criado, M. (2006). Alkali activated fly ash binders. A comparative study between sodium and potassium activators. *Materiales de Construción, 56(281)*, 51–65. https://doi.org/10.3989/mc.2006.v56.i281.92.

Fořt, J., Vejmelková, E., Keppert, M., Rovnaníková, P., Bezdička, P. & Černý, R. (2020). Alkaline activation of low-reactivity ceramics: Peculiarities induced by the precursors' dual character. *Cement and Concrete Composites, 105*, 103440. https://doi.org/10.1016/j.cemconcomp.2019.103440.

Frías, M., Rodríguez, O., Vegas, I. & Vigil, R. (2008). Properties of calcined clay waste and its influence on blended cement behavior. *Journal of the American Ceramic Society, 91(4)*, 1226-1230. https://doi.org/10.1111/j.1551-2916.2008.02289.x.

Gaibor, N., Coelho, J., Leitão, D., Miranda, T., Tavares, P. & Cristelo, N. (2020). Alkali activation of recycled ceramic aggregates from construction and demolition wastes. *Materiales de Construcción, 70(339)*, e222-e222. http://digital.casalini.it/4947315.

Gaibor, N., Leitão, D., Miranda, T., Cristelo, N., Fernandes, L., Pereira, E. N. & Cunha, V. M. (2023). Fiber-reinforced alkali-activated cements from ceramic waste and ladle furnace slag without thermal curing. *Journal of Materials in Civil Engineering, 35(9)*, 04023271. https://doi.org/10.1061/JMCEE7.MTENG-14776.

Garcia-Lodeiro, I., Palomo, A. & Fernández-Jiménez, A. (2015). An overview of the chemistry of alkali-activated cement-based binders. In *Handbook of Alkali-activated Cements, Mortars and Concretes* (pp. 19–47). Elsevier Ltd. https://doi.org/10.1533/9781782422884.1.19.

García-Lodeiro, I., Donatello, S., Fernández-Jiménez, A. & Palomo, A. (2016). Hydration of hybrid alkaline cement containing a very large proportion of fly ash: A descriptive model. *Materials, 9*, 605. https://doi.org/10.3390/ma9070605.

Gartner, E. (2004). Industrially Interesting Approaches to "Low-CO_2" Cement. *Cement and Concretes Research, 34*, 1489-1498. http://dx.doi.org/10.1016/j.cemconres.2004.01.021.

Gautam, L., Jain, J. K., Kalla, P. & Choudhary, S. (2021). A review on the utilization of ceramic waste in sustainable construction products. *Materials Today: Proceedings, 43*, 1884-1891. https://doi.org/10.1016/j.matpr.2020.10.829.

Gavali, H. R., Bras, A., Faria, P. & Ralegaonkar, R. V. (2019). Development of sustainable alkali-activated bricks using industrial wastes. *Construction and Building Materials, 215*, 180-191. https://doi.org/10.1016/j.conbuildmat.2019.04.152.

Glukhovsky, V. D. (1967). Soil silicate articles and structures, in Russian, (Gruntosilikatnye vyroby I konstruktsii), Ed. Budivelnyk Publisher., Kiev.

Heller-Kallai, L. (2006). Thermally modified clay minerals. *Developments in clay science, 1*, 289-308. https://doi.org/10.1016/S1572-4352(05)01009-3.

Heller-Kallai, L. (2013). Thermally modified clay minerals. In Developments in clay science (Vol. 5, pp. 411-433). Elsevier. https://doi.org/10.1016/B978-0-08-098258-8.00014-6.

Hwang, C. L., Yehualaw, M. D., Vo, D. H. & Huynh, T. P. (2019). Development of high-strength alkali-activated pastes containing high volumes of waste brick and ceramic powders. *Construction and Building Materials, 218*, 519-529. https://doi.org/10.1016/j.conbuildmat.2019.05.143.

Hou, D., Sun, M., Wang, M., Wan, X., Chen, Z., Wang, X., Zhang, Y. & Wang, P. (2023). Molecular insight into the formation and fracture process of sodium aluminosilicate hydrate gels. *The Journal of Physical Chemistry C, 127(31)*, 15542-15555. https://doi.org/10.1021/acs.jpcc.3c02632.

Huseien, G. F., Sam, A. R. M., Shah, K. W., Asaad, M. A., Tahir, M. M. & Mirza, J. (2019). Properties of ceramic tile waste based alkali-activated mortars incorporating GBFS and fly ash. *Construction and Building Materials, 214*, 355-368. https://doi.org/10.1016/j.conbuildmat.2019.04.154.

Huertas, F. J., Fiore, S. & Linares, J. (1997). Thermal analysis as a tool for determining and defining spherical kaolinite. *Clays and clay minerals, 45*, 587-590. https://doi.org/10.1346/CCMN.1997.0450410.

Hong, F., Wang, M., Dong, B., Diao, X., Zhang, X., Pang, K. Zhang, Y. & Hou, D. (2023). Molecular Insight into the Pozzolanic Reaction of Metakaolin and Calcium Hydroxide. *Langmuir, 39(10)*, 3601-3609. https://doi.org/10.1021/acs.langmuir.2c03115.

Jain, P., Gupta, R. & Chaudhary, S. (2022). A literature review on the effect of using ceramic waste as supplementary cementitious material in cement composites on workability and compressive strength. *Materials Today: Proceedings, 65*, 871-876. https://doi.org/10.1016/j.matpr.2022.03.453.

Jiménez, J. R., Ayuso, J., López, M., Fernández, J. M. & De Brito, J. M. C. L. (2013). Use of fine recycled aggregates from ceramic waste in masonry mortar manufacturing. *Construction and Building Materials, 40*, 679-690. https://doi.org/10.1016/j.conbuildmat.2012.11.036.

Juan-Valdes, A., Medina, C., Morán, J. M., Guerra, M. I., Aguado, P. J., De Rojas, M. I. S., Frías, M. & Rodriguez, O. (2010). Re-Use of Ceramic Wastes in Construction. In *Ceramic Materials* (pp. 197-214). Intechopen. https://doi.org/10.5772/intechopen.83933.

Katzer, J. (2013). Strength performance comparison of mortars made with waste fine aggregate and ceramic fume. *Construction and Building Materials, 47*, 1-6. https://doi.org/10.1016/j.conbuildmat.2013.04.039.

Keppert, M., Vejmelková, E., Bezdička, P., Doleželová, M., Čáchová, M., Scheinherrová, L., Pokorný, J., Vyšvařil, M., Rovnaníková, P. & Černý, R. (2018). Red-clay ceramic powders as geopolymer precursors: Consideration of amorphous portion and CaO content. *Applied Clay Science, 161*, 82-89. https://doi.org/10.1016/j.clay.2018.04.019.

Komnitsas, K. A, Zaharaki, D. & Perdikatsis, V. (2007). Geopolymerisation of low calcium ferronickel slags. *Journal of Materials Science, 42*, pp. 3073-3082. https://doi.org/10.1007/s10853-006-0529-2.

Komnitsas, K. A. (2011). Potential of geopolymer technology towards green buildings and sustainable cities. *Procedia Engineering, 21*, 1023–1032. https://doi.org/10.1016/j.proeng.2011.11.2108.

Koyuncu, H., Guney, Y., Yılmaz, G., Koyuncu, S. & Bakis, R. (2004). Utilization of ceramic wastes in the construction sector. *Key Engineering Materials, 264*, 2509-2512. https://doi.org/10.4028/www.scientific.net/KEM.264-268.2509.

Krivenko, P. (2017). Why alkaline activation – 60 years of the theory and practice of alkali-activated materials. *Journal of Ceramic Science and Technology, 08(3)*, 323-334. https://doi.org/10.4416/JCST2017-00042.

Larrosa, C., Lotero, A., Bastos, C., Servi, S.; & Consoli, N. C. (2023). Stabilization of dredging material from Rio Grande harbor with alkali-activated cements to produce masonry elements. *Journal of Materials in Civil Engineering, 35(4)*, 04023018. https://doi.org/10.1061/(ASCE)MT.1943-5533.0004677.

Lotero, A., Consoli, N. C., Moncaleano, C. J., Tebechrani Neto, A & Koester, E. (2021). Mechanical properties of alkali-activated ground waste glass-carbide lime blends for geotechnical uses. *Journal of Materials in Civil Engineering, 33(10)*, 04021284. https://doi.org/10.1061/(ASCE)MT.1943-5533.0003918.

Lotero, A, Moncaleano, C. J. & Consoli. N.C. (2023). Alkali-activated red ceramic wastes-carbide lime blend: An alternative alkaline cement manufactured at room temperature. *Journal of Building Engineering, 65(2023)*, 105663. https://doi.org/10.1016/j.jobe.2022.105663.

Luhar, I., Luhar, S., Abdullah, M. M. A. B., Nabiałek, M., Sandu, A.V., Szmidla, J., Jurczyńska, A., Razak, R. A., Aziz, I. H. A., Jamil, N.H. & Deraman, L. M. (2021). Assessment of the Suitability of Ceramic Waste in *Geopolymer Composites: An Appraisal. Materials, 14(12)*, 3279. https://doi.org/10.3390/ma14123279.

Luukkonen, T., Abdollahnejad, Z., Yliniemi, J., Kinnunen, P. & Illikainen, M. (2018). Comparison of alkali and silica sources in one-part alkali-activated blast furnace slag mortar. *Journal of Cleaner Production, 187*, 171-179. https://doi.org/10.1016/j.jclepro.2018.03.202.

McLellan, B. C., Williams, R. P., Lay, J., Van Riessen, A. & Corder, G. D. (2011). Costs and carbon emissions for geopolymer pastes in comparison to ordinary Portland cement. *Journal of Cleaner Production, 19*, 1080-1090. https://doi.org/10.1016/j.jclepro.2011.02.010.

Magbool, H. M. (2022). Utilisation of ceramic waste aggregate and its effect on eco-friendly concrete: A review. *Journal of Building Engineering, 47*, 103815. https://doi.org/10.1016/j.jobe.2021.103815.

Mbereyaho, L., Amini, F., Bigirimana, F. & Isengwe, R. C. (2019). The use of ceramic tiles waste in concrete as partial replacement of fine aggregate. *Mediterranean Journal of Basic and Applied Sciences, 3(3)*, 122-130.

Medina, M., Rojas, C. & Moisés, M. F. (2012). Reuse of sanitary ceramic wastes as coarse aggregate in eco-efficient concretes. *Cement and Concrete Composites, 34*, 48–54. https://doi.org/10.1016/j.cemconcomp.2011.08.015.

Mikuni, A., Komatsu, R. & Ikeda, K. (2007). Dissolution properties of some fly ash fillers applying to geopolymeric materials in alkali solution. *Journal of Materials Science, 42*, 2953–2957. https://doi.org/10.1007/s10853-006-0530-9.

Mohana, R. & Bharathi, S. L. (2023). Sustainability of pre-treated and nano-fly ash powder on the thermal stability and environmental impact of green mortars under ambient conditions. *Journal of Building Engineering, 71*, 106494. https://doi.org/10.1016/j.jobe.2023.106494.

Núñez, V., Lotero, A., Bastos, C. A., Sargent, P., Consoli, N. C. (2023). Mechanical and microstructure analysis of mass-stabilized organic clay thermally cured using a ternary binder. *Acta Geotech*. https://doi.org/10.1007/s11440-023-01961-x.

Pacheco-Torgal, F., Castro-Gomes, J. & Jalali, S. (2008). Alkali-activated binders: A review Part 1. Historical background, terminology, reaction mechanisms and hydration products. *Construction and Building Materials, 22*, 1305–1314. https://doi.org/10.1016/j.conbuildmat.2007.10.015.

Pacheco-Torgal, F. & Jalali, S. (2010). Reusing ceramic wastes in concrete. *Construction and Building* Materials, 24(5), 832–838. https://doi.org/10.1016/j.conbuildmat.2009.10.023.

Palomo, A., Maltseva, O., Garcia-Lodeiro, I. & Fernández-Jimenez, A. (2013). Hybrid alkaline cements: Part II. Clinker factor. *Romanian Journal of Materials*, 43, 74–79.

Palomo, A., Krivenko, P., Garcia-Lodeiro, I., Kavalerova, E., Maltseva, O. & Fernández-Jiménez, A. (2014). A review on alkaline activation: new analytical perspectives. *Materiales de Construcción, 64(315)*, e022. https://doi.org/10.3989/mc.2014.00314.

Provis J. & Van Deventer J.S.J. (eds) (2009), *Geopolymers: Structure, Processing, Properties and Industrial Aplications* (1st ed.), Woodhead Publishing Limited, Cambridge.

Provis, J. L. (2018). Alkali-activated materials. *Cement and Concrete Research, 114*, 40-48. https://doi.org/10.1016/j.cemconres.2017.02.009.

Puertas, F. (1995). Alkaline activated slag cements: current situation and future perspectives. *Materials and Construction, 45(239)*, 53-64. http://dx.doi.org/10.3989/mc.1995.v45.i239.553.

Puertas, F., Barba, A., Gazulla, M., Gómez, M., Palacios, M. & Martínez, S. (2006). Ceramic wastes as raw materials in Portland cement clinker fabrication: characterization and alkaline activation. *Materiales de Construcción, 56(281)*, 73–84. https://doi.org/10.3989/mc.2006.v56.i281.94.

Puertas, F., Garcia-Diaz, I., Barba, A., Gazulla, M., Palacios, M. & Gomez, M. (2008). Ceramic wastes as alternative raw materials for Portland cement clinker production. *Cement and Concrete Composites, 30(9)*, 798–805. https://doi.org/10.1016/j.cemconcomp.2008.06.003.

Puertas, F., Palacios, M., Manzano, H., Dolado, J. S., Rico, A. & Rodríguez, J. (2011). A model for the C-A-S-H gel formed in alkali-activated slag cements. *Journal of the European Ceramic Society, 31(12)*, 2043–2056. https://doi.org/10.1016/j.jeurceramsoc.2011.04.036.

Rakhimova, N. R. & Rakhimov, R. Z. (2015). Alkali-activated cements and mortars based on blast furnace slag and red clay brick waste. *Materials & design, 85*, 324-331. https://doi.org/10.1016/j.matdes.2015.06.182.

Rakhimova, N. R., Morozov, V. P. & Eskin. A. A. (2021). Phase formation behavior of alkali-activated calcined clays: effects of the reactive phase and NaOH concentration. *Geosystem Engineering, 24 (5-6)*, 238-246. https://doi.org/10.1080/12269328.2020.1768154.

Rakhimova, N. R. (2022). Recent advances in blended alkali-activated cements: A review. *European Journal of Environmental and Civil Engineering, 26(10)*, 4596-4618. https://doi.org/10.1080/19648189.2020.1858170.

Ray, S., Haque, M., Sakib, M. N., Mita, A. F., Rahman, M. M. & Tanmoy, B. B. (2021). Use of ceramic wastes as aggregates in concrete production: A review. *Journal of Building Engineering, 43*, 102567. https://doi.org/10.1016/j.jobe.2021.102567.

Reig, L., Tashima, M. M., Borrachero, M. V., Monzó, J., Cheeseman, C. R. & Payá, J. (2013a). Properties and microstructure of alkali-activated red clay brick waste. *Construction and Building Materials, 43*, 98–106. https://doi.org/10.1016/j.conbuildmat.2013.01.031.

Reig, L., Tashima, M. M., Soriano, L., Borrachero, M. V., Monzó, J. & Payá, J. (2013b). Alkaline activation of ceramic waste materials. *Waste Biomass Valorization, 4*, 729–736. https://doi.org/10.1007/s12649-013-9197-z.

Reig, L., Borrachero, M. V., Monzó, J. M., Savastano, H., Tashima, M. M. & Payá, J. (2016). Use of ceramic sanitaryware as an alternative for the development of new sustainable binders. *Key Engineering Materials, 668*, 172-180. https://doi.org/10.4028/www.scientific.net/KEM.668.172.

Reig, L., Sanz, M. A., Borrachero, M. V., Monzó, J., Soriano, L. & Payá, J. (2017). Compressive strength and microstructure of alkali-activated mortars with high ceramic waste content. *Ceramics International, 43(16)*, 13622-13634. https://doi.org/10.1016/j.ceramint.2017.07.072.

Ricciotti, L., Occhicone, A., Ferone, C., Cioffi, R., Tarallo, O. & Roviello, G. (2022). Development of geopolymer-based materials with ceramic waste for artistic and restoration applications. *Materials, 15(23)*, 8600. https://doi.org/10.3390/ma15238600.

Rivera, J. F., Aguirre, A. M. G., Gutiérrez, R. M. & Orobio, A. (2020). Chemical stabilization of soils - conventional and alkali-activated materials. *Informador Técnico, 84 (2)*. http://doi.org/10.23850/22565035.2530.

Robayo, R.A., Mulford, A., Munera, J. & Mejía de Gutiérrez, R. (2016). Alternative cements based on alkali-activated red clay brick waste. *Construction and Building Materials, 128*, 163–169. https://doi.org/10.1016/j.conbuildmat.2016.10.023.

Ruiz-Santaquiteria, C.; Fernández-Jiménez, A.; Skibsted, J.; Palomo, A. (2013). Clay reactivity: Production of alkali activated cements. Applied Clay Science, 73, 11–16. https://doi.org/10.1016/j.clay.2012.10.012.

Samadi, M., Huseien, G. F., Mohammadhosseini, H., Lee, H. S., Lim, N. H. A. S., Tahir, M. M. & Alyousef, R. (2020). Waste ceramic as low cost and eco-friendly materials in the production of sustainable mortars. *Journal of Cleaner Production, 266*, 121825. https://doi.org/10.1016/j.jclepro.2020.121825.

Saldanha, R. B., Caicedo, A. M. L., De Araújo, M. T., Scheuermann Filho, H. C., Moncaleano, C. J., Silva, J. P. S., & Consoli, N. C. (2023). Potential use of iron ore tailings for binder production: A life cycle assessment. *Construction and Building Materials*, 365, 130008. https://doi.org/10.1016/j.conbuildmat.2022.130008.

Salustio, J., Torres, S. M., Melo, A. C., Silva, Â. J. C., Azevedo, A. C., Tavares, J. C., Leal, M. S. & Delgado, J. M. P. Q. (2022). Mortar bond strength: A brief literature review, tests for analysis, new research needs and initial experiments. *Materials, 15*, 2332. https://doi.org/10.3390/ma15062332.

Sarkar, M. & Dana, K. (2021). Partial replacement of metakaolin with red ceramic waste in geopolymer. *Ceramics International, 47(3)*, 3473-3483. https://doi.org/10.1016/j.ceramint.2020.09.191.

Servi, S., Lotero, A., Silva, J. P. S., Bastos, C & Consoli, N. C. (2022). Mechanical response of filtered and compacted iron ore tailings with different cementing agents: Focus on tailings-binder mixtures disposal by stacking. *Construction and Building Materials, 349*, 128770. https://doi.org/10.1016/j.conbuildmat.2022.128770.

Shi, C., Roy, D. M. & Krivenko, P. (2003). *Alkali-activated cements and concretes* (1st ed.). Taylor & Francis, Abingdon, UK.

Shi, C., Fernández-Jiménez, A. & Palomo, A. (2011). New cements for the 21st century: The pursuit of an alternative to Portland cement. *Cement and Concrete Research, 41*, 750-763. https://doi.org/10.1016/j.cemconres.2011.03.016.

Silva, T. H., Castro, A. C. M., Valente, F. C., Soares, M. M. N. S., De Resende, D. S. & Bezerra, A. C. S. (2019). Recycling ceramic waste as a raw material in sanitary ware production. *Cerâmica, 65*, 426-431. https://doi.org/10.1590/0366-69132019653752687.

Sperinck, S., Raiteri, P., Marks, N. & Wright, K. (2011). Dehydroxylation of kaolinite to metakaolin—a molecular dynamics study. *Journal of Materials Chemistry, 21(7)*, 2118-2125. https://doi.org/10.1039/C0JM01748E.

Sudo, T., Shimoda, S., Nishigaki, S. & Aoki, M. (1967). Energy changes in dehydration processes of clay minerals. *Clay Minerals, 7(1)*, 33-42. https://doi.org/10.1180/claymin.1967.007.1.03.

Sun, Z., Cui, H., An, H., Tao, D., Xu, Y., Zhai, J. & Li, Q. (2013). Synthesis and thermal behavior of geopolymer-type material from waste ceramic. *Construction and Building Materials, 49*, 281-287. https://doi.org/10.1016/j.conbuildmat.2013.08.063.

Torres-Carrasco, M., Palomo, J. G. & Puertas, F. (2014). Sodium silicate solutions from dissolution of glasswastes. Statistical analysis. *Materiales de Construcción, 64(314)*, e014-e014. http://digital.casalini.it/2960701.

Tuyan, M., Andiç-Çakir, Ö. & Ramyar, K. (2018). Effect of alkali activator concentration and curing condition on strength and microstructure of waste clay brick powder-based geopolymer. *Composites Part B: Engineering, 135*, 242-252. https://doi.org/10.1016/j.compositesb.2017.10.013.

Van Deventer, J. S. J., Provis, J. L., Duxson, P. & Lukey, G. C. (2007). Reaction mechanisms in the geopolymeric conversion of inorganic waste to useful products *Journal of Hazardous Materials, 139*, 506-513. https://doi.org/10.1016/j.jhazmat.2006.02.044.

Van Deventer, J. S. J., Provis, J. L., Duxson, P. & Brice, D. G. (2010). Chemical research and climate change as drivers in the commercial adoption of alkali activated materials. *Waste and Biomass Valorization, 1*, 145–155. https://doi.org/10.1007/S12649-010-9015-9.

Wachtman, J. B., Cannon, W. R. & Matthewson, M. J. (2009). *Mechanical Properties of Ceramics*. (2nd ed.). Wiley.

Wang, S. & Scrivener, K. L. (1995). Hydration products of alkali activated slag cement. *Cement and Concrete Research, 25*, 561–571. https://doi.org/10.1016/0008-8846(95)00045-E.

Wang, S. D., Pu, X. C., Scrivener, K. L. & Pratt, P. L. (1995). Alkali-activated slag: A review of properties and problems. *Cement and Concrete Research, l7*, 93–102. https://doi.org/10.1680/adcr.1995.7.27.93.

Weise, K., Ukrainczyk, N. & Koenders, E. (2023). Pozzolanic reactions of metakaolin with calcium hydroxide: Review on hydrate phase formations and effect of alkali hydroxides, carbonates and sulfates. *Materials & Design, 231*, 112062. https://doi.org/10.1016/j.matdes.2023.112062.

Wu, Y., Lu, B., Bai, T., Wang, H., Du, F., Zhang, Y., Cai, L., Jiang, C. & Wang, W. (2019). Geopolymer, green alkali activated cementitious material: Synthesis, applications and challenges. *Construction and Building Materials, 224*, 930-949. https://doi.org/10.1016/j.conbuildmat.2019.07.112.

Xiao, J., Li, W., Fan, Y. & Huang, X. (2012). An overview of study on recycled aggregate concrete in China (1996–2011). *Construction and building materials, 31*, 364-383. https://doi.org/10.1016/j.conbuildmat.2011.12.074.

Yang, K. H., Mun, J. H., Sim, J. I. & Song, J. K. (2011). Effect of water content on the properties of lightweight alkali-activated slag concrete. *Journal of materials in civil engineering, 23(6)*, 886-894. https://doi.org/10.1061/(ASCE)MT.1943-5533.0000244.

Zawrah, M. F., Gado, R. A., Feltin, N., Ducourtieux, S. & Devoille, L. J. P. S. (2016). Recycling and utilization assessment of waste fired clay bricks (Grog) with granulated blast-furnace slag for geopolymer production. *Process Safety and Environmental Protection, 103*, 237-251. https://doi.org/10.1016/j.psep.2016.08.001.

Zhu, X., Qian, H., Wu, H., Zhou, Q., Feng, H., Zeng, Q., Tian, Y., Ruan, S., Zhang, Y., Chen, S. & Yan, D. (2022). Early-stage geopolymerization process of metakaolin-based geopolymer. *Materials, 15(17)*, 6125. https://doi.org/10.3390/ma15176125.

Chapter 8

Alkali-Activated Materials Applied to Reclaimed Asphalt Pavement Stabilization: A Literature Review

Deise Trevizan Pelissaro[1,*]
and Francisco Dalla Rosa[2]
[1]Department of Civil Engineering, Integrated Regional University of Alto Uruguay and Missões, Erechim, RS, Brazil
[2]Graduate Program in Civil and Environmental Engineering, University of Passo Fundo, Passo Fundo, RS, Brazil

Abstract

From a social and environmental point of view, the use of industrial and construction waste is an alternative to the preservation and reuse of natural resources. In paving, the use of reclaimed asphalt pavement (RAP) in pavement recycling is an alternative for the structural and functional recovery of the highway. Despite the technical, economic, and environmental advantages of the method, RAP often needs to be stabilized by a material with cementing properties, usually Portland cement. However, due to the high levels of CO_2 emitted during their production, alkali-activated materials present themselves as an alternative to replacing traditional cementitious materials. The alkali-activation process consists of mixing an aluminosilicate mineral (precursor) and a highly alkaline solution (activator). This systematic review discussed the parameters evaluated in several studies and the properties of RAP mixtures in alkali-activated cementitious matrices, as well as changes in microstructure. The biggest challenge in stabilizing RAP through alkali-activated systems is the small number of studies still

* Corresponding Author's Email: deisetrevizan@uricer.edu.br.

In: Environmental Geotechnics
Editors: E. Pavan Korf, P. Domingos Marques Prietto, C. Müller et al.
ISBN: 979-8-89113-651-9

available, mainly related to the assessment of the environmental, social, and economic impacts of this technique. Literature research also indicated the need to develop alternative activators that use silica sources from agricultural or industrial waste in order to enable the practical application of these materials. Despite the lack of research evidence, RAP stabilization through alkali-activated systems is a promising solution for pavement base and subbase layers.

Keywords: alkali-activation, reclaimed asphalt pavement, systematic review, geotechnical stabilization

Introduction

Alkali-activated materials are a new alternative to Portland cement, mainly regarding CO_2 emissions into the environment. These binders are the product of the reaction between an aluminosilicate mineral (precursor) and an alkaline solution (activator) (Provis & Bernal, 2014; Provis & Deventer, 2014). The name alkali-activated products is given to systems produced with materials with the presence of Al_2O_3, SiO_2, and a high content of CaO, while materials with a reduced content of CaO are called geopolymers. Although there is this difference in nomenclature, the two groups of materials are often used without distinction, generating disagreements in the literature.

This class of materials is very versatile due to the wide variety of raw materials and production methods. More than 65% of the Earth's crust consists of Al-Si-based materials and a wide range of these materials can serve as a source for the synthesis of alkali-activated materials (Xu & Van Deventer, 2000). Numerous industrial byproducts, agricultural residues, and other waste products can be used as a source of aluminosilicate materials, such as fly ash, granulated blast furnace slag, rice husk ash, and palm oil ash (Davidovits, 2008; Duxson et al., 2007; Yang & Song, 2009). The first studies and applications of alkali-activated materials mainly used precursors such as blast furnace slag, fly ash, and metakaolin (Majidi, 2009; Provis & Deventer, 2014; Shi et al., 2006). However, currently alkali-activated materials are also an effective way to recycle and reuse various industry by-products through the inclusion of, for example, fly ash (Aydın & Baradan, 2012; Giasuddin et al., 2013; Oh et al., 2010), clay-based slag (Ranjbar et al., 2015; Yusuf et al., 2014), rice husk ash (He et al., 2013), to make them more sustainable in terms of environment, availability, and economy. The fact that these precursor materials are produced by industries around the world makes alkali-activated materials sustainable in terms of availability. 780 million tons/year of fly ash are produced worldwide (Duan et al., 2016), 11 million tons/year of palm oil ash (Vakili et al., 2015), 20 million tons/year of rice husk ash (Kamseu et al., 2017; Sinulingga et al., 2014), 120 million tons/year of red alumina mud (Duan et al., 2016; Ye et al., 2016), in addition to the existence of huge deposits of kaolin (Liew et al., 2016; Zhang et al., 2016).

Furthermore, aluminosilicate materials do not react with water and therefore require aqueous solutions with a high pH, identified as a precursor and usually produced from a mixture of hydroxides, silicates, and water (Longhi et al., 2019). The most commonly used alkaline activators are sodium silicate, sodium hydroxide, sodium carbonate, potassium hydroxide, and potassium silicate (Rashad et al., 2013). Regarding the alkaline activator, Duxson et al. (2007) state that its main objective is to supply alkali metal cations, in order to increase the pH of the mixture and accelerate the dissolution of the solid precursor. A highly alkaline medium is essential for the appropriate activation of the silicon and aluminum present in the precursor material, enabling the transformation of a partially or completely glassy structure into a very compact material (Eduok, 2016). In the economic sphere, to reduce high maintenance costs, researchers and managers have been seeking to find new options for using materials, reducing energy consumption, speeding up restoration interventions, among others. While in relation to the environmental sector, the aim is to minimize the consumption of natural aggregates, earth movements, fuels used, among others.

Currently, sustainability actions in pavements include the use of different types of recycled aggregates, the most common being recycled concrete (Kisku et al., 2017), slag (Huang et al., 2007), waste glass (Jamshidi et al., 2016), crushed brick (Mohammadinia et al., 2016a), and reclaimed asphalt pavement (RAP) (Milad et al., 2020). RAP is a waste produced during pavement rehabilitation and a very large amount of this material is still not recycled and is discarded in landfills. Some studies claim that recycling reclaimed asphalt pavement can reduce the overall cost of pavement sections by 40%-46% (Debbarma et al., 2019; Singh et al., 2018). Furthermore, a resolution was recently published in Brazil, which states that all restoration, capacity adjustment, and expansion works of the National Department of Transport Infrastructure (DNIT) must include the reuse of the RAP.

The implementation of RAP as a pavement construction material stands out when it comes to base layers, as they are thicker and, therefore, can consume greater volumes of recycled materials. However, the use of large quantities of recycled aggregates generally represents a loss of properties that needs to be compensated with the use of binders or stabilizers, that is, agents that increase the compressive and impact resistance and the general integrity of the aggregates layers (Arulrajah et al., 2014; Avirneni et al., 2016). For chemical stabilization, materials with binding characteristics have been used, such as hydrated lime or Portland cement (OPC), being widely used for layers such as soil-lime and soil-cement (Ingles & Metcalf, 1972). The most used stabilizer for base layers is Portland cement, a material with a large contribution to the emission of greenhouse gases, which are released during its manufacture. Each ton of cement produces approximately 0.87 tons of carbon dioxide, which is responsible for approximately 6%-7% of anthropogenic CO_2 emissions, in addition to a significant reduction in natural resources (Aprianti et al., 2015).

In parallel with the need to minimize energy consumption and greenhouse gas emissions into the atmosphere, there is a growing concern regarding the storage and disposal of waste materials or industrial by-products. Therefore, one of the possible alternatives to Portland cement are alkali-activated materials, which can reduce the carbon footprint by up to 80% (Duxson et al., 2007) and still have good mechanical properties and durability.

Reclaimed Asphalt Pavement (RAP)

Reclaimed asphalt pavement removed from an existing pavement is a fully recyclable material (AAPA, 2018; EAPA, 2014). It is composed of two valuable non-renewable resources, i.e., approximately 95% by weight of aggregates and 5% by weight of aged bituminous binder. This material can be reused in new asphalt mixtures, reducing the demand for virgin aggregates and bitumen, or as recycled aggregates to produce non-bonded layers of pavements (SABITA, 2019). This composition makes the reclaimed asphalt pavement at the top of the hierarchy of recyclable materials with engineering applications, as it is not reduced as simply aggregated (SABITA, 2019). As a result, it can, significantly reduce the global costs of new bituminous products (Williams et al., 2020).

Reclaimed asphalt pavement has been used in asphalt mixes since 1915 for environmental and economic reasons (Kennedy et al., 1998). Currently, reclaimed asphalt pavement represents one of the most reused construction products worldwide. In 2018, approximately 88 wt% and 72 wt% of reclaimed asphalt pavement were used in the US and Europe, respectively, as aggregates for hot, hot, and cold asphalt mixes and for unbound layers (Tarsi et al., 2020).

In Brazil, road paving works saw a huge increase in the 1950s, when, as a result of the intense exchange of technicians from the extinct DNER, there was a large transfer of technology from the United States of North America (DNIT, 2006). However, the use of reclaimed asphalt pavement was particularly attractive during the 1970s, especially with the rise in oil prices during the Arab Oil Restriction. As a consequence, asphalt paving technologists have been forced to develop recycling methods that are capable of reducing the demand for asphalt binder, thus reducing the cost of asphalt paving mixtures (Milad et al., 2020).

In addition, a resolution recently came into force in Brazil, which states that all engineering projects for the restoration, capacity adaptation, and expansion of the National Department of Transport Infrastructure (DNIT) must include the reuse of the RAP eventually produced in the enterprise.

According to Babashamsi et al. (2016), a life cycle cost analysis of pavements incorporating reclaimed asphalt pavement showed that savings of \$58,000/km can be achieved on asphalt mixtures containing 30% to 50% of this material. This savings is due to the replacement of part of the virgin aggregates and the lower transportation cost. Conventional crushed stone is generally more expensive than recycled materials. In 2014, the National Asphalt Pavement Association (NAPA) reported that approximately 71.9 million tons of reclaimed asphalt pavement were used in the construction of new pavements in the United States. This means savings of around US\$2.5 billion each year (George et al., 2021). Additionally, in 2018, more than 100 million tons of reclaimed asphalt pavement were collected for reuse in the United States, saving an estimated 61.4 million cubic yards from landfills (Song et al., 2018). Therefore, the recycling and/or reuse of reclaimed asphalt pavement is an example of sustainable development in the infrastructure sector. The encouragement of its use reflects the global tendency to face existing environmental issues, seeking to increase the efficient use of resources and reduce carbon emissions (United Nations, 2015).

Despite the importance and advantages of recycling pavements, the biggest challenge in using reclaimed asphalt pavement is due to the thin layer of aged bitumen on the surface of the aggregate (Arulrajah et al., 2014; Saride et al., 2016). The asphalt layer on the surface of the aggregate is responsible for the low water absorption of the material and the weak bond between the aggregate and the cement mortar (Brand & Roesler, 2017; Saride et al., 2016; Singh et al., 2018; Su et al., 2014). According to Saride et al. (2016), processing techniques can help in the partial or total removal of the asphalt coating, but they are considered expensive and unfeasible (Singh et al., 2018).

Age can also play an important role in the aggregate properties of reclaimed asphalt pavement. Singh et al. (2018) observed that newer pavement aggregates had lower workability and worse properties than aggregates originating from older pavements. Brand & Roesler (2017) suggest that material that is older or stored for a long time in piles undergoes chemical oxidation that modifies the chemistry of the surface asphalt, leading to better adhesion of the cement paste.

In this way, the performance of reclaimed asphalt pavement can be improved by mixing with virgin aggregates and/or adding rejuvenating agents, new asphalt binders, foamed asphalt, asphalt emulsions, and even Portland cement or other chemical additives (Bernucci et al., 2008). In this sense, concerns related to the strength of recycled waste materials have led to extensive investigation into the stabilization of recycled aggregates using a variety of binders such as Portland cement, lime, slag, calcium carbide residue, and fly ash (Gnanendran & Piratheepan, 2010; Horpibulsuk et al., 2015; Puppala et al., 2011; Saride et al., 2015).

Alkali-Activated Materials as Binders

Alkali-activated materials (AAM) are increasingly used and widely studied worldwide (Liu et al., 2019; Provis & Bernal, 2014; Shi et al., 2011). These materials were developed as an alternative to replace Portland cement, used in concrete, mortars, and pastes, and are considered less aggressive from an environmental point of view, due to lower atmospheric CO_2 emissions when compared to the production of Portland cement (Kua et al., 2019; Zhuang et al., 2016). According to Davidovits (2008), alkali-activated materials are the product of a reaction between an alkaline activator and an aluminosilicate source. Depending on the Al and Ca content, AAM can be further classified as inorganic polymers or geopolymers, or simply as low-calcium (low-Ca) and high-calcium (high-Ca) AAM (Samarakoon et al., 2019) (see Chapter 4).

The raw materials used are generally industrial by-products, with little or no attributed environmental impact and low cost. In addition to being rich in alumina (Al_2O_3) and silica (SiO_2), the precursor materials, both in their natural form and in the form of industrial waste or other raw material with low added value, must preferably have reactive amorphous characteristics. These aluminosilicate precursors require alkaline activation solutions that generally consist of an aqueous solution of alkaline hydroxides, silicates, carbonates, or sulfates (Alanazi et al., 2016; De Vargas et al., 2011; Hu et al., 2009).

Zhang et al. (2013) classify precursor materials as calcined (fly ash, slag, construction waste, and pozzolanic waste) and non-calcined (kaolinite, feldspar, clays, and mineral processing waste). The authors explain that the calcination process increases the reactivity of the materials, by changing the crystalline structure into an amorphous one. Therefore, calcined precursors present a faster dissolution rate and, therefore, greater resistance to initial compression when compared to non-calcined precursors, which present more significant increases in mechanical resistance in the stages after the alkali-activation reaction.

More than 65% of the Earth's crust consists of Al-Si-based materials and a wide range of these materials can serve as a source for the synthesis of alkali-activated materials (Xu & Van Deventer, 2000). The first studies and applications of alkali-activated materials mainly used precursors such as blast furnace slag, fly ash, and metakaolin (Majidi, 2009; Provis & Deventer, 2014; Shi et al., 2006). However, currently alkali-activated materials are also an effective way to recycle and reuse various industry by-products through the inclusion of, for example, fly ash (Aydın & Baradan, 2012; Giasuddin et al., 2013; Oh et al., 2010), clay-based slag (Ranjbar et al., 2015; Yusuf et al., 2014), rice husk ash (He et al., 2013), to make them more sustainable in terms of environment, availability, and economy.

The fact that these precursor materials are produced by industries around the world makes alkali-activated materials sustainable also in terms of availability. 780 million tons/year of fly ash are produced worldwide (Bakri et al., 2013; Duan et al., 2016), 11 million tons/year of palm oil ash (Vakili et al., 2015), 20 million tons/year of rice husk ash (Kamseu et al., 2017; Sinulingga et al., 2014), 120 million tons/year of red alumina mud (Duan et al., 2016; Ye et al., 2016), in addition to the existence of huge deposits of kaolin (Liew et al., 2016; Zhang et al., 2016).

Regarding the alkaline activator, Duxson et al. (2007) state that its main objective is to supply alkali metal cations, to increase the pH of the mixture and accelerate the dissolution of the solid precursor. A highly alkaline medium is essential for the appropriate activation of the silicon and aluminum present in the precursor material, enabling the transformation of a partially or completely glassy structure into a very compact material (Eduok, 2016).

Alkali-activated materials are a very versatile class of materials due to the wide variety of raw materials, forms of production, and potential reduction in CO_2 emissions compared to the use of Portland cement-based products (Provis, 2018). According to Turner & Collins (2013), for each kilogram of cement processed, around 0.66 kg to 0.82 kg of CO_2 is released.

Manufacturing concrete from geopolymers can result in a reduction in CO_2 emissions of 26% to 45% compared to the production of Portland cement. Furthermore, they may have properties and characteristics such as high mechanical strength, fire resistance, low shrinkage, and low thermal conductivity (Duxson et al., 2007; Ma et al., 2018). These are the reaction mechanisms that determine the behavior of a material whose formation comes from a process of chemical reactions (Pacheco-Torgal et al., 2015).

Alkali-Activation Applied to Stabilization of Reclaimed Asphalt Pavement

Over the years, several studies have evaluated the potential for incorporating reclaimed asphalt pavement into the base and sub-base layers of pavement. However, it is observed that for high substitution levels, there are reductions in mechanical properties, which can be compensated by the use of stabilizers. It is possible to use materials such as lime, cement, fly ash, blast furnace slag, rice husk ash, and sugarcane bagasse ash, among others (Singh et al., 2018). Alkali-activated materials have emerged in the last two decades as sustainable construction materials replacing Portland cement and more recently these new binders have been studied as stabilizers in base and sub-base layers in combination with reclaimed asphalt pavement (Costa et al., 2020). The literature on the subject is still limited and Table 1 presents some studies highlighting the precursors and activators used, as well as the proportions of virgin aggregates (VA), the replacement percentages of reclaimed asphalt pavement, the curing temperature, the parameters evaluated, and the range of unconfined compressive strength (UCS) values obtained in them.

Through the results presented in Table 1, it can first be observed that the main precursor used is fly ash, covering 78.6% of the works evaluated. This aspect highlights the high potential of this residue performing the function of activated alkali precursor. In the same context, in the case of another essential element for the development of alkali-activated materials, activators composed of the combination of NaOH and Na_2SiO_3 are used in approximately 80% of the research carried out.

In this context, Hoy et al. (2016a) evaluated the stabilization of RAP with alkaline-activated fly ash using a solution of sodium hydroxide and sodium silicate in different proportions. The results show that the resistance increases as the $NaOH/Na_2SiO_3$ ratio decreases both at room temperature and for curing at 40°C. Microscopic analyzes indicate low geopolymerization when only NaOH ($NaOH/Na_2SiO_3$ = 100:0) is used as an alkaline solution, but with increasing curing time and temperature there is greater dissolution of the silica and alumina present in the fly ash, increasing the resistance.

A major environmental concern when using waste and alkali-activated materials is the possibility of soil contamination and waterways due to leaching. The durability of base layers may also be affected by leaching of stabilizers during service life (Costa et al., 2020). By this way, as a complement to the work of Hoy et al. (2016a, 2016b) evaluated the leaching capacity of heavy metals for the stabilization of reclaimed asphalt pavement with both alkali-activated fly ash and fly ash alone. The results showed that there is no environmental risk for both mixtures and that the geopolymeric binder reduces the leaching of heavy metals, further increasing the advantages in the application of recycled materials in a sustainable way in road construction. Horpibulsuk et al. (2017) also reached the same conclusions, identifying that after six wetting and drying cycles, resistance begins to decrease.

It is also observed that the works in general use high levels of RAP, with the aggregate matrix even being composed of 100% RAP in some works. However, some studies carried out on RAP-OPC systems detected that the presence of the asphalt layer is detrimental to the connection of RAP to the matrix. In other words, the RAP particles showed better adhesion to the matrix for exposed aggregate surfaces (without asphalt layer). In that regard Saride et al. (2016) investigated the use of fly ash (20% and 30%) with and without the activation of a hydroxide solution in proportions of 2% and 4%. The great difference of the work is that the results are presented depending on the exposed surface area of the reclaimed asphalt pavement, since the aggregates are covered with a layer of amorphous asphalt that influences the formation of pozzolanic compounds.

Jallu et al. (2020) investigated the flexural fatigue behavior of RAP stabilized with alkali-activated fly ash. The highest values of compressive strength and resilience modulus were obtained for the combination Na_2SiO_3:NaOH equal to 70:30. Samples of beams reinforced with biaxial and triaxial geogrids were also evaluated, where the beams reinforced with the triaxial geogrid had a fatigue life 1.66 times greater than those reinforced with a biaxial geogrid and 2.54 times greater than the control samples. The authors identified great potential in stabilizing reclaimed asphalt pavement from a fly ash-based geopolymer as a pavement base material.

Table 1. Summary of studies of stabilization reclaimed asphalt pavement from alkali-activated materials

Reference	Precursors	% precursor	Activators	NaOH: Na_2SiO_3	Molarity	% RAP	%VA	Resistance (MPa)	Cure temperature	Evaluated parameters
Avimeni et al. (2016)	Fly ash	20; 30	NaOH	-	1; 2 M	80; 60	20; 40	1.5–5.5	RT	UCS, durability, microstructure, leaching
Hoy et al. (2016a)	Fly ash	10; 20; 30	NaOH; Na_2SiO_3	100:0; 90:10; 60:40; 50:50	10 M	100	-	1.9–7.0	RT; 40°C	UCS, microstructure
Hoy et al. (2016b)	Fly ash	20	NaOH; Na_2SiO_3	100:0; 90:10; 60:40; 50:50	10 M	100	-	1.9–5.6	RT	UCS, microstructure, leaching
Hoy et al. (2018)	Slag	10; 20	NaOH; Na_2SiO_3	100:0; 70:30; 60:40; 50:50	8 M	100	-	0.75–3.0	RT	UCS, microstructure
Mohammadinia et al. (2016a)	Fly ash; Slag	2; 4	NaOH; Na_2SiO_3	-	8 M	100	-	1.0–2.2	RT	UCS, elasticity and resilience modulus
Mohammadinia et al. (2016b)	Fly ash	4; 8;16	NaOH; Na_2SiO_3	-	8 M	100	-	1.0–2.2	RT; 40°C	UCS, resilience modulus
Saride et al. (2016)	Fly ash	20; 30	NaOH	-	1; 2 M	80; 60	20; 40	1.5–6.5	RT	UCS, microstructure
Horpibulsuk et al. (2017)	Fly ash	20; 30	NaOH; Na_2SiO_3	100:0; 90:10; 60:40; 50:50	10 M	100	-	1.9–6.0	RT	UCS, durability, microstructure, leaching
Jallu et al. (2020)	Fly ash	20	NaOH; Na_2SiO_3	70:30; 50:50	3 M	60	40	1.5–6.5	RT	Flexion fatigue
Saride e Jallu (2020)	Fly ash	20	NaOH; Na_2SiO_3	100:0; 70:30; 50:50; 30:70; 10:90	0.5; 1; 3 M	60	40	0.2–6.5	27°C	UCS, resilience modulus, durability

Reference	Precursors	% precursor	Activators	NaOH: Na_2SiO_3	Molarity	% RAP	%VA	Resistance (MPa)	Cure temperature	Evaluated parameters
Widayanti et al. (2021)	Fly ash	5	NaOH; Na_2SiO_3	2.5	8 M	45; 55; 65; 75	-	1.0 – 3.5	RT	Marshall; microstructure
Goldoni et al. (2023)	Fly ash	10; 20; 30	NaOH; Na_2SiO_3	80:20; 90:10; 70:30	10 M	20; 45;70	80; 55; 30	0.4 – 32.0	RT	UCS, durability
Pelissaro et al. (2023)	Metakaolin	20:30:40	Rice hush ask; NaOH	-	10 M	30; 40; 50	70; 60; 50	1.0 - 12.0	RT; 60°C	UCS; microstructure

*RT = room temperature (23 ± 2°C).

Mohammadinia et al. (2016a) compared the performance of three recycled construction and demolition materials: crushed brick, recycled crushed aggregate, and RAP stabilized with geopolymers produced from fly ash and ground granulated blast furnace slag. Pozzolanic binders were used separately at a content of 4% and together at equal levels of 2% and were compared with Portland cement. Both the resilience modulus and the compressive strength increased with the use of the geopolymer for all materials, however for the recycled crushed aggregate the stabilization was more effective. The slag-based geopolymer proved to be more efficient than the fly ash-based one. All mixtures of reclaimed asphalt pavement with geopolymer proved to be sufficient for application in highway sub-bases at 7 days, however, only the use of 4% slag geopolymer at 28 days provided the requirements for the pavement base. Slag-based geopolymers were also studied by Hoy et al. (2018) who showed that it is essential to use Na_2SiO_3 for application on highway bases after 7 days. However, the authors emphasize that the use of excess Na_2SiO_3 delayed the rate of geopolymerization, leading to a reduction in resistance.

Mohammadinia et al. (2016b) also evaluated the strength and resilience modulus for the same construction and demolition waste but for only the fly ash geopolymer in an alkaline environment of sodium silicate and sodium hydroxide. The results show that the use of geopolymers considerably increases the resilience modulus, the compressive strength increases with the increase in the amount of ash, and that the heat treatment increases the reaction rate and consequently the resistance as well. Finally, the authors state that it is possible to use recycled crushed aggregate and reclaimed asphalt pavement with the addition of 8% and 16% respectively of fly ash for the pavement base for curing conditions at 40°C and 7 days.

Adhikari et al. (2020) proposed a regression model to evaluate the sensitivity of the variables involved in the stabilization of two soils with fly ash geopolymer and RAP. Higher contents of fly ash and reclaimed asphalt pavement produced mixtures with higher strengths. The ideal sodium silicate content depends on the type of soil and the percentage of fly ash and reclaimed asphalt pavement. The mechanical and durability properties of the optimal mixture were better than or similar to conventional soil-cement mixtures.

Curing conditions also appear to have a significant impact on mechanical properties as they increase phase dissociation (reactivity). Alkali-activated RAP samples cured for seven days at 40°C showed approximately 70% higher strength than reference samples cured at room temperature (Mohammadinia et al., 2016b). However, high-temperature curing is not accessible for practical applications. Furthermore, subjecting the samples to higher curing temperatures tends to avoid the phenomenon of efflorescence as can be observed in the work of Pelissaro et al. (2023). However, the authors state that this increase in temperature can also result in cracks which, in turn, impair the mechanical behavior of the specimens over time. This work is also promising in the use of an alternative silicate source, bringing promising results and highlighting the importance of reducing the environmental and economic impacts of using alkali-activated materials in RAP stabilization.

Finally, it is observed that in addition to the amount of research on RAP-AAM being limited; most studies evaluated short healing periods. However, recently the work of Goldoni et al. (2023) evaluated RAP-AAM matrices for periods of up to 360 days, showing that the long-term did not show a significant reduction in resistance over time, which indicates that the binder remained in the RAP mixtures. Thus, alkali-activated materials can be an alternative to traditional Portland cement. The authors also carried out durability tests and although the specimens were subjected to weathering cycles, all mixtures presented UCS superior to the standard requirements for paving applications. However, the authors state that the durability test methodology was not useful for evaluating the degradation of the mixtures, since resistance increased over the cycles. This evidence suggests that weathering cycles played a role in curing due to the drying process in the thermal chamber (Goldoni et al., 2023).

Final Considerations

The aforementioned studies reveal the possibility of applying the alkali-activation process in the recycling of reclaimed asphalt pavement. However, the amount of research on stabilization of reclaimed asphalt pavement from alkali-activation is still limited and most studies used only fly ash as a precursor, and little is known about the mechanical and durability properties. Furthermore, existing studies are based on the use of sodium hydroxide and/or silicate, highlighting considerable improvements in the performance of final products due to the action of silicate. Despite this, sodium silicate has a high energy consumption and considerable CO_2 emissions, which justifies the use of local materials that may include industrial and agricultural waste.

In this context, it is necessary to produce an alternative silicate, which presents economic and environmental benefits and enables the long-term application of alkali- activated materials on a large scale, as is the case in the infrastructure area. Although there are already studies regarding the development of alternative silicate based on the dissolution of rice hush ash, there is no extensive knowledge about the complete production process.

The use of alkali-activation in the stabilization of reclaimed asphalt pavement is very promising in the construction of base and sub-base layers of pavements, due to the increase in resistance provided by alkali-activated binders and the low environmental impact when alternative solutions to commercial sodium silicate are used. In this sense, there is a large gap in knowledge regarding alkali-activation using alternative sodium silicate. As in the research already carried out, it is necessary to understand the mechanical and microstructural behavior for different proportions of aggregates and activating alkali solution, as well as the evaluation of the durability and alkali leaching properties. Another important factor to be considered is the curing healing conditions, which until now have not been the main objective of the works mentioned in this item.

References

AAPA. (2018). Reclaimed asphalt pavement (RAP) management plan. *National Technology & Leadership Committee*.

Adhikari, S., Khattak, M. J., & Adhikari, B. (2020). Mechanical characteristics of soil-RAP-geopolymer mixtures for road base and subbase layers. *International Journal of Pavement Engineering*, *21*(4), 483–496. https://doi.org/10.1080/10298436.2018.1492131.

Alanazi, H., Yang, M., Zhang, D. & Gao, Z. (2016). Bond strength of PCC pavement repairs using metakaolin-based geopolymer mortar. *Cement and Concrete Composites*, *65*, 75–82. https://doi.org/10.1016/j.cemconcomp.2015.10.009.

Aprianti, E., Shafigh, P., Bahri, S. & Farahani, J. N. (2015). Supplementary cementitious materials origin from agricultural wastes - A review. *Construction and Building Materials*, *74*, 176–187. https://doi.org/10.1016/j.conbuildmat.2014.10.010.

Arulrajah, A., Piratheepan, J. & Disfani, M. M. (2014). Reclaimed asphalt pavement and recycled concrete aggregate blends in pavement subbases: Laboratory and field evaluation. *Journal of Materials in Civil Engineering*, *26*(2), 349–357. https://doi.org/10.1061/(asce)mt.1943-5533.0000850.

Avirneni, D., Peddinti, P. R. T. & Saride, S. (2016). Durability and long term performance of geopolymer stabilized reclaimed asphalt pavement base courses. *Construction and Building Materials*, *121*, 198–209. https://doi.org/10.1016/j.conbuildmat.2016.05.162.

Aydın, S. & Baradan, B. (2012). Mechanical and microstructural properties of heat cured alkali-activated slag mortars. *Materials and Design*, *35*, 374–383.

Babashamsi, P., Md Yusoff, N. I., Ceylan, H., Md Nor, N. G. & Salarzadeh Jenatabadi, H. (2016). Evaluation of pavement life cycle cost analysis: Review and analysis. *International Journal of Pavement Research and Technology, 9(4),* 241-254. https://doi.org/10.1016/j.ijprt.2016.08.004.

Bakri, A. M. M. Al, Abdulkareem, O. A., Rafiza, A. R., Zarina, Y., Norazian, M. N. & Kamarudin, H. (2013). Review on processing of low calcium fly ash geopolymer concrete. *Australian Journal of Basic and Applied Sciences*, *7*(5), 342–349.

Bernucci, L. B., Motta, L. M. G., Ceratti, J. A. P. & Soares, J. B. (2008). *Pavimentação Asfáltica: Formação Básica para Engenheiros. Rio de Janeiro. PETROBRAS: ABEDA.*

Brand, A. S. & Roesler, J. R. (2017). Bonding in cementitious materials with asphalt-coated particles: Part I – The interfacial transition zone. *Construction and Building Materials*, *130*, 171–181. https://doi.org/10.1016/j.conbuildmat.2016.10.019.

Costa, J. O., Borges, P. H. R., dos Santos, F. A., Bezerra, A. C. S., Van den bergh, W., & Blom, J. (2020). Cementitious binders and reclaimed asphalt aggregates for sustainable pavement base layers: Potential, challenges and research needs. *Construction and Building Materials*, *265*, 120325. https://doi.org/10.1016/j.conbuildmat.2020.120325.

Davidovits, J. (2008). *Geopolymer Chemistry and Applications*. (J. Davidovits (ed.); 2nd ed.). Institut Géopolymère.

De Vargas, A. S., Dal Molin, D. C. C., Vilela, A. C. F., Silva, F. J. Da, Pavão, B. & Veit, H. (2011). The effects of Na_2O/SiO_2 molar ratio, curing temperature and age on compressive strength, morphology and microstructure of alkali-activated fly ash-based geopolymers. *Cement and Concrete Composites*, *33*(6), 653–660. https://doi.org/10.1016/j.cemconcomp.2011.03.006.

Debbarma, S., Ransinchung, G. D. & Singh, S. (2019). Feasibility of roller compacted concrete pavement containing different fractions of reclaimed asphalt pavement. *Construction and Building Materials*, *199*, 508–525. https://doi.org/10.1016/j.conbuildmat.2018.12.047.

DNIT. (2006). *Manual de restauração de pavimentos asfálticos*. (DNIT (ed.); 2th ed.). IPR 720.

Duan, P., Yan, C. & Zhou, W. (2016). Influence of partial replacement of fly ash by metakaolin on mechanical properties and microstructure of fly ash geopolymer paste exposed to sulfate attack. *Ceramics International*, *42*(2), 3504–3517. https://doi.org/10.1016/j.ceramint.2015.10.154.

Duxson, P., Fernández-Jiménez, A., Provis, J. L., Lukey, G. C., Palomo, A. & Van Deventer, J. S. J. (2007). Geopolymer technology: The current state of the art. *Journal of Materials Science*, *42*(9), 2917–2933. https://doi.org/10.1007/s10853-006-0637-z.

EAPA. (2014). Asphalt the 100% Recyclable Construction Product. *EAPA Position Paper*: Brussels, Belgium.

Eduok, E. I. (2016). *Thermal Properties of Geopolymer Materials*. University of Stavanger.

George, A. M., Banerjee, A., Puppala, A. J. & Saladhi, M. (2021). Performance evaluation of geocell-reinforced reclaimed asphalt pavement (RAP) bases in flexible pavements. *International Journal of Pavement Engineering*, *22*(2), 181–191. https://doi.org/10.1080/10298436.2019.1587437.

Giasuddin, H. M., Sanjayan, J. G. & P.G, R. (2013). Strength of geopolymer cured in saline water in ambient conditions. *Fuel*, *107*, 34–39. https://doi.org/10.1016/j.fuel.2013.01.035.

Gnanendran, C. T. & Piratheepan, J. (2010). Determination of fatigue life of a granular base material lightly stabilized with slag lime from indirect diametral tensile testing. *Journal of Transportation Engineering*, *136*(8), 736–745. https://doi.org/10.1061/(ASCE)TE.1943-5436.0000138.

Goldoni, A. G., Pelissaro, D. T., Silveira, E., Prietto, P. D. M. & Rosa, F. D. (2023). Durability and mechanical long-term performance of reclaimed asphalt pavement stabilized by alkali-activation. *Soils and Rocks*, *46*(1), 1–9. https://doi.org/10.28927/SR.2023.007422.

He, J., Jie, Y., Zhang, J., Yu, Y. & Zhang, G. (2013). Synthesis and characterization of red mud and rice husk ash-based geopolymer composites. *Cement and Concrete Composites*, *37*(1), 108–118. https://doi.org/10.1016/j.cemconcomp.2012.11.010.

Horpibulsuk, S., Hoy, M., Witchayaphong, P., Rachan, R. & Arulrajah, A. (2017). Recycled asphalt pavement – fly ash geopolymer as a sustainable stabilized pavement material. *IOP Conference Series: Materials Science and Engineering*, *273*, 012005. https://doi.org/10.1088/1757-899x/273/1/012005.

Horpibulsuk, S., Kampala, A., Phetchuay, C., Udomchai, A. & Arulrajah, A. (2015). Calcium carbide residue - A cementing agent for sustainable soil stabilization. *Geotechnical Engineering Journal of the SEAGS and AGSSEA*, *46*, 22–27.

Hoy, M., Horpibulsuk, S., & Arulrajah, A. (2016a). Strength development of Recycled Asphalt Pavement - fly ash geopolymer as a road construction material. *Construction and Building Materials*, *117*, 209–219. https://doi.org/10.1016/j.conbuildmat.2016.04.136.

Hoy, M., Horpibulsuk, S., Arulrajah, A. & Mohajerani, A. (2018). Strength and microstructural study of recycled asphalt pavement: Slag geopolymer as a pavement base material. *Journal of Materials in Civil Engineering*, *30*(8), 1–11. https://doi.org/10.1061/(asce)mt.1943-5533.0002393.

Hoy, M., Horpibulsuk, S., Rachan, R., Chinkulkijniwat, A. & Arulrajah, A. (2016b). Recycled asphalt pavement – fly ash geopolymers as a sustainable pavement base material: Strength and toxic leaching investigations. *Science of the Total Environment*, *573*, 19–26. https://doi.org/10.1016/j.scitotenv.2016.08.078.

Hu, M., Zhu, X. & Long, F. (2009). Alkali-activated fly ash-based geopolymers with zeolite or bentonite as additives. *Cement and Concrete Composites*, *31*(10), 762–768. https://doi.org/10.1016/j.cemconcomp.2009.07.006.

Huang, Y., Bird, R. N. & Heidrich, O. (2007). A review of the use of recycled solid waste materials in asphalt pavements. *Resources, Conservation and Recycling*, *52*(1), 58–73. https://doi.org/10.1016/j.resconrec.2007.02.002.

Ingles, O. G. & Metcalf, J. B. (1972). *Soil Stabilization – Principles and Practice*. (1st ed.) Ed. Butterworths, Sydney.

Jallu, M., Arulrajah, A., Saride, S. & Evans, R. (2020). Flexural fatigue behavior of fly ash geopolymer stabilized-geogrid reinforced RAP bases. *Construction and Building Materials*, *254*, 119263. https://doi.org/10.1016/j.conbuildmat.2020.119263.

Jamshidi, A., Kurumisawa, K., Nawa, T. & Igarashi, T. (2016). Performance of pavements incorporating waste glass: The current state of the art. *Renewable and Sustainable Energy Reviews*, *64*, 211–236. https://doi.org/10.1016/j.rser.2016.06.012.

Kamseu, E., Beleuk à Moungam, L. M., Cannio, M., Billong, N., Chaysuwan, D., Melo, U. C. & Leonelli, C. (2017). Substitution of sodium silicate with rice husk ash-NaOH solution in metakaolin based geopolymer cement concerning reduction in global warming. *Journal of Cleaner Production*, *142*, 3050–3060. https://doi.org/10.1016/j.jclepro.2016.10.164.

Kennedy, T. W., Tam, W. O. & Solaimanian, M. (1998). Optimizing use of reclaimed asphalt pavement with the superpave system. *Journal of the Association of Asphalt Paving Technologists*, *67*, 311–333.

Kisku, N., Joshi, H., Ansari, M., Panda, S. K., Nayak, S., & Dutta, S. C. (2017). A critical review and assessment for usage of recycled aggregate as sustainable construction material. *Construction and Building Materials*, *131*, 721–740. https://doi.org/10.1016/j.conbuildmat.2016.11.029.

Kua, T. A., Imteaz, M. A., Arulrajah, A. & Horpibulsuk, S. (2019). Environmental and economic viability of alkali activated material (AAM) comprising slag, fly ash and spent coffee ground. *International Journal of Sustainable Engineering*, *12*(4), 223–232. https://doi.org/10.1080/19397038.2018.1492043.

Liew, Y. M., Heah, C. Y., Mustafa, A. B. M. & Kamarudin, H. (2016). Structure and properties of clay-based geopolymer cements: A review. *Progress in Materials Science*, *83*, 595–629.

Liu, Y., Shi, C., Zhang, Z. & Li, N. (2019). An overview on the reuse of waste glasses in alkali-activated materials. *Resources, Conservation and Recycling*, *144*, 297–309. https://doi.org/10.1016/j.resconrec.2019.02.007.

Longhi, M. A., Walkley, B., Rodríguez, E. D., Kirchheim, A. P., Zhang, Z. & Wang, H. (2019). New selective dissolution process to quantify reaction extent and product stability in metakaolin-based geopolymers. *Composites Part B: Engineering*, *176*. https://doi.org/10.1016/j.compositesb.2019.107172.

Ma, C. K., Awang, A. Z. & Omar, W. (2018). Structural and material performance of geopolymer concrete: A review. *Construction and Building Materials*, *186*, 90–102. https://doi.org/10.1016/j.conbuildmat.2018.07.111.

Majidi, B. (2009). Geopolymer technology, from fundamentals to advanced applications: A review. *Materials Technology*, *24*, 79–87. https://doi.org/10.1179/175355509X449355.

Milad, A., Taib, A. M., Ahmeda, A. G. F., Solla, M. & Yusoff, N. I. M. (2020). A review of the use of reclaimed asphalt pavement for road paving applications. *Jurnal Teknologi*, *82*(3), 35–45. https://doi.org/10.11113/jt.v82.14320.

Mohammadinia, A., Arulrajah, A., Sanjayan, J., Disfani, M. M., Bo, M. W. & Darmawan, S. (2016a). Strength development and microfabric structure of construction and demolition aggregates stabilized with fly ash–based geopolymers. *Journal of Materials in Civil Engineering*, *28*(11), 1–8. https://doi.org/10.1061/(asce)mt.1943-5533.0001652.

Mohammadinia, A., Arulrajah, A., Sanjayan, J., Disfani, M. M., Win Bo, M. & Darmawan, S. (2016b). Stabilization of demolition materials for pavement base/subbase applications using fly ash and slag geopolymers: Laboratory investigation. *Journal of Materials in Civil Engineering*, *28*(7). https://doi.org/10.1061/(asce)mt.1943-5533.0001526.

Oh, J. E., Monteiro, P. J. M., Jun, S. S., Choi, S. & Clark, S. M. (2010). The evolution of strength and crystalline phases for alkali-activated ground blast furnace slag and fly ash-based geopolymers. *Cement and Concrete Research*, *40*, 189–196. https://doi.org/10.1016/j.cemconres.2009.10.010.

Pacheco-Torgal, F., Labrincha, J., Leonelli, C., Palomo, A. & Chindaprasi, P. (2015). *Handbook of Alkali-Activated Cements, Mortars and Concretes* (F. Pacheco-Torgal, J. A. Labrincha, C. Leonelli, A. Palomo, P. Chindaprasirt (eds.); 2th ed.). Woodhead Publishing.

Pelissaro, D. T., Bruschi, G. J., Korf, E. P. & Rosa, F. D. (2023). Rice husk ash as an alternative soluble silica source for alkali-activated metakaolin systems applied to recycled asphalt pavement stabilization. *Transportation Geotechnics*, *39*, 100940. https://doi.org/10.1016/j.trgeo.2023.100940.

Provis, J. L. (2018). Alkali-activated materials. *Cement and Concrete Research*, *114*, 40–48. https://doi.org/10.1016/j.cemconres.2017.02.009.

Provis, J. L. & Bernal, S. A. (2014). Geopolymers and related alkali-activated materials. *Annual Review of Materials Research*, *44*, 299–327. https://doi.org/10.1146/annurev-matsci-070813-113515.

Provis, J. L., & Deventer, J. S. J. van. (2014). *Alkali-Activated Materials: State-of-the-Art Report* (J. L. Provis, J. S. J. van Deventer (eds); 1th ed.). Springer.

Puppala, A. J., Hoyos, L. R. & Potturi, A. K. (2011). Resilient moduli response of moderately cement-treated reclaimed asphalt pavement aggregates. *Journal of Materials in Civil Engineering*, *23*(7). https://doi.org/10.1061/(ASCE)MT.1943-5533.000026.

Ranjbar, N., Mehrali, M., Mehrali, M., Alengaram, U. J. & Jumaat, M. Z. (2015). Graphene nanoplatelet-fly ash based geopolymer composites. *Cement and Concrete Research*, 76, 222–231. https://doi.org/10.1016/j.cemconres.2015.06.003.

Rashad, A. M., Bai, Y., Basheer, P. A. M., Milestone, N. B. & Collier, N. C. (2013). Hydration and properties of sodium sulfate activated slag. *Cement and Concrete Composites*, *37*(1), 20–29. https://doi.org/10.1016/j.cemconcomp.2012.12.010.

SABITA, S. A. B. A. (2019). Use of reclaimed asphalt in the production of asphalt. *Manual36/Trh 21:Western Cape, South Africa*.

Samarakoon, M. H., Ranjith, P. G., Rathnaweera, T. D. & Perera, M. S. A. (2019). Recent advances in alkaline cement binders: A review. *Journal of Cleaner Production*, *227*, 70–87. https://doi.org/10.1016/j.jclepro.2019.04.103.

Saride, S., Avirneni, D., & Challapalli, S. (2016). Micro-mechanical interaction of activated fly ash mortar and reclaimed asphalt pavement materials. *Construction and Building Materials*, *123*, 424–435. https://doi.org/10.1016/j.conbuildmat.2016.07.016.

Saride, S., Avirneni, D., Javvadi, S. C. P., Puppala, A. J. & Hoyos, L. R. (2015). Evaluation of fly ash treated reclaimed asphalt pavement for base/subbase applications. *Indian Geotechnical Journal*, *45*(4), 401–411. https://doi.org/10.1007/s40098-014-0137-z.

Shi, C., Jiménez, A. F. & Palomo, A. (2011). New cements for the 21st century: The pursuit of an alternative to Portland cement. *Cement and Concrete Research*, *41*(7), 750–763. https://doi.org/10.1016/j.cemconres.2011.03.016.

Shi, C., Roy, D. & Krivenko, P. (2006). *Alkali-Activated Cements and Concretes*; 1st ed. Taylor & Francis Group.

Singh, S., Ransinchung, G. D. R. N., Monu, K. & Kumar, P. (2018). Laboratory investigation of RAP aggregates for dry lean concrete mixes. *Construction and Building Materials*, *166*, 808–816. https://doi.org/10.1016/j.conbuildmat.2018.01.131.

Sinulingga, K., Agusnar, H., Wirjosentono, B. & Amin, Z. M. (2014). The effect of mixing rice husk ash and palm oil boiler ash on concrete strength. *American Journal of Physical Chemistry*, *3*(2), 9–14. 10.11648/j.ajpc.20140302.11.

Song, W., Huang, B. & Shu, X. (2018). Influence of warm-mix asphalt technology and rejuvenator on performance of asphalt mixtures containing 50% reclaimed asphalt pavement. *Journal of Cleaner Production*, *192*, 191–198. https://doi.org/10.1016/j.jclepro.2018.04.269.

Su, Y.-M., Hossiney, N., Tia, M. & Bergin, M. (2014). Mechanical properties assessment of concrete containing reclaimed asphalt pavement using the superpave indirect tensile strength test. *Journal of Testing and Evaluation*, *42*(4), 912–920. 10.1520/JTE20130093.

Tarsi, G., Tataranni, P. & Sangiorgi, C. (2020). The challenges of using reclaimed asphalt pavement for new asphalt mixtures: A review. *Materials*, *13*(18). https://doi.org/10.3390/ma13184052.

Turner, L. K. & Collins, F. G. (2013). Carbon dioxide equivalent (CO_2-e) emissions: A comparison between geopolymer and OPC cement concrete. *Construction and Building Materials*, *43*, 125–130. https://doi.org/10.1016/j.conbuildmat.2013.01.023.

United Nations. (2015). Transforming our world: The 2030 agenda for sustainable development. *United Nations: New York, NY, USA*.

Vakili, M., Rafatullah, M., Ibrahim, M. H., Salamatinia, B., Gholami, Z. & Zwain, H. M. (2015). A review on composting of oil palm biomass. *Environment, Development and Sustainability*, *17*(4), 691–709. https://doi.org/10.1007/s10668-014-9581-2.

Widayanti, A., Soemitro, R., Ekaputri, J. J. & Suprayitno, H. (2021). Asphalt concrete mixture produced using reclaimed asphalt pavement and fly ash as artificial aggregate and filler. *Jurnal Teknologi*, *83*(4), 17–29. https://doi.org/10.11113/jurnalteknologi.v83.16289.

Williams, B. A., Willis, J. R. & Shacat, J. (2020). Annual asphalt pavement industry survey on recycled materials and warm-mix asphalt usage: 2019. *National Asphalt Pavement Association*, 48.

Xu, H. & Van Deventer, J. S. J. (2000). The geopolymerization of alumino-silicate mínerals. *International Journal of Mineral Processing*, *59*, 247–266.

Yang, K. H. & Song, J. K. (2009). Workability loss and compressive strength development of cementless mortars activated by combination of sodium silicate and sodium hydroxide. *Journal of Materials in Civil Engineering*, *21*(3), 119–127. https://doi.org/10.1061/(asce)0899-1561(2009)21:3(119).

Ye, N., Chen, Y., Yang, J., Liang, S., Hu, Y., Xiao, B., Huang, Q., Shi, Y., Hu, J. & Wu, X. (2016). Co-disposal of MSWI fly ash and Bayer red mud using an one-part geopolymeric system. *Journal of Hazardous Materials*, *318*, 70–78. https://doi.org/10.1016/j.jhazmat.2016.06.042.

Yusuf, M. O., Johari, M. A. M., Maslehuddin, M. & Ahmad, Z. A. (2014). Influence of curing methods and concentration of NaOH on strength of the synthesized alkaline activated ground slag-ultrafine palm oil fuel ash mortar/concrete. *Construction and Building Materials*, *66*, 541–548. https://doi.org/10.1016/j.conbuildmat.2014.05.037.

Zhang, M., Guo, H., El-Korchi, T., Zhang, G. & Tao, M. (2013). Experimental feasibility study of geopolymer as the next-generation soil stabilizer. *Construction and Building Materials*, *47*, 1468–1478. https://doi.org/10.1016/j.conbuildmat.2013.06.017.

Zhang, Z. H., Zhu, H. J., Zhou, C. H. & Wang, H. (2016). Geopolymer from kaolin in China: An overview. *Applied Clay Science*, *119*, 31–41. https://doi.org/10.1016/j.clay.2015.04.023.

Zhuang, X. Y., Chen, L., Komarneni, S., Zhou, C. H., Tong, D. S., Yang, H. M., Yu, W. H. & Wang, H. (2016). Fly ash-based geopolymer: Clean production, properties and applications. *Journal of Cleaner Production*, *125*, 253–267. https://doi.org/10.1016/j.jclepro.2016.03.019.

Chapter 9

Efficacy of Alkali-Activated Fly Ash-Based Deep Soil Mixing Column to Stabilize Expansive Clay

Ammavajjala Sesha Sai Raghuram[1]
Rousseau Prasanna[2,3]
Anasua Guharay[1]
and Priyadharshini Perumal[4,*]

[1]Department of Civil Engineering, Birla Institute of Technology and Science, Pilani-Hyderabad, Shamirpet, India
[2]School of Civil and Environmental Engineering, Indian Institute of Technology Mandi, Himachal Pradesh, India
[3]Department of Civil and Environmental Engineering, Carleton University, Ottawa, Canada
[4]Fibre and Particle Engineering Research Unit, University of Oulu, Oulu, Finland

Abstract

This study examines the efficacy of alkali-activated fly ash (AAFA) for stabilizing expansive clays (EC) through the in-situ deep soil mixing (DSM) technique. The main objective of the study is to mitigate the swell-shrink behavior of EC and attain improved strength properties through AAFA. Due to the very limited calcium oxide content of Class-F fly ash (FA), an alkaline environment is necessary to generate pozzolanic reactions involving silica and alumina. The alkali environment within the FA was established through the introduction of an alkali activator, which consisted of a mixture of sodium silicate, and

* Corresponding Author's Email: Priyadharshini.perumal@oulu.fi.

In: Environmental Geotechnics
Editors: E. Pavan Korf, P. Domingos Marques Prietto, C. Müller et al.
ISBN: 979-8-89113-651-9

sodium hydroxide. Small-scale laboratory tests were conducted on the EC stabilized with DSM columns utilizing AAFA. The experiments were carried out by varying the l/d ratio of the DSM column. The outcomes of the investigation indicated that the surface heave characteristics of the EC have reduced significantly with increasing the l/d of the DSM column. The percentage reduction in heave increased from 59.45% to 75.67% when the l/d ratio of the DSM column increased from 1.66 to 3.33. Furthermore, the DSM column demonstrates an efficient reduction of surface heave at a radial distance of 1.5 times the diameter of the DSM column from its center. Moreover, a considerable increase in the ultimate load-bearing capacity has been observed. The results indicated that for a given settlement of 20 mm, the load-carrying capacity was increased from 185 kPa to 1293 kPa when the DSM column of l/d = 3.33.

Keywords: expansive soils, alkali-activated fly ash, heave, deep soil mixing

Introduction

The expansive soils, known for their swelling and shrinkage characteristics, are distributed throughout the world, especially in arid and semi-arid regions. In India, Black cotton soil, a type of expansive soil, is abundantly found in the states of Maharashtra, Madhya Pradesh, Gujarat, Karnataka, Andhra Pradesh, and Chhattisgarh (Mir 2002). This soil poses a serious threat to the infrastructure facilities built upon them due to their swelling or shrinkage behaviour. The swelling-shrinkage characteristics of expansive soil are mainly attributed to the presence of clay minerals from the smectite group, such as Montmorillonite, Bentonite, and Illite (Holtz et al., 2011). In these clay minerals, the ions are held together by weak van der Waals forces, which give them a strong affinity for water. With an increase in the soil's moisture content, these clay minerals in the expansive soils undergo extensive volumetric deformation resulting in swelling. The developed swelling pressure can cause damage to foundations, underground utility lines and pavements. Upon drying, the expansive soils undergo shrinkage, leading to settlement in foundations and rutting of pavements. Each year, this annual swelling and shrinkage behaviour of expansive soils incurs significant costs associated with the damage and repair of infrastructure facilities around the world (Considine, 1984; Dang et al., 2015; Jones & Jefferson, 2012; Li et al., 2014; Sidiq et al., 2023). Therefore, it is highly essential to stabilize expansive soils through mechanical or chemical methods to ensure the safe functioning of infrastructure facilities (see Chapter 3).

The mechanical stabilization of expansive soil involves improving the soil without the addition of any stabilizing agents. These methods include prewetting, compaction, moisture barriers, soil replacement, and soil reinforcement (Petry & Amstrong, 1989; Zada et al., 2023). Despite several advantages of these methods, including ease of application, cost-effectiveness, and reduced environmental impact, they do have notable drawbacks. Issues such as inhomogeneity in soil treatment and the necessity for supplementation by chemical methods are significant drawbacks (Zada et al. 2023). To address these shortcomings and consider the specific needs of a project, chemical stabilization methods are often favored over mechanical methods for stabilizing expansive soils. The chemical method of stabilization, as described by (Petry & Amstrong, 1989), involves modifying the physicochemical properties of the clay by mixing it with cementitious binding agents. This process changes the interaction of clay minerals with water, thereby affecting the overall engineering behavior of the soil mass. This chemical stabilization is employed within deep soil mixing (DSM), which is a common *in-situ* method of stabilizing expansive clays. It involves mechanical blending of soil with cementitious binding slurry to the targeted stabilization depth using specialized drilling and mixing tools (FHWA, 2013; Mypati & Saride, 2022). Commonly, within DSM, cement and lime serve as primary binding agents for stabilizing expansive clays.

Numerous studies have investigated the effectiveness of lime, cement, or their combination in reducing the swelling-shrinkage characteristics and improving shear strength properties of expansive soils (Al-Tabbaa & Evans, 1999; Barman & Dash, 2022; Bhadriraju et al., 2008; Eujine et al., 2017; Sherwood, 1993; Lorenzo & Bergado, 2006; Madhyannapu & Puppala, 2014; Raja & Thyagaraj, 2020; Sahoo & Pradhan, 2010; Saride et al., 2013; Thyagaraj et al., 2012). In recent years, due to environmental concerns regarding greenhouse gas emissions from the manufacturing of traditional binders such as cement and lime, there has been an increasing trend towards using low-carbon binders derived from waste materials sourced from power plants, mines, and agriculture for stabilizing expansive soils (Mypati & Saride, 2022; Syed et al. 2023). Some of the eco-friendly binding materials that could be a viable alternative to cement and lime are fly ash (FA), pond ash, ground granulated blast furnace slag (GGBS), rice husk ash, and sugarcane bagasse ash etc. (Jain et al., 2020; Okabue & Onyeobi, 1999; Mypati & Saride, 2022; Syed et al., 2020, 2023; Taha et al., 2021; Tanyıldızı et al., 2023; Xu & Yi,

2021; Zada et al., 2023). Notably, each of these eco-friendly binders possesses distinct advantages and limitations. As a result, further extensive research is necessary to assess the suitability of each of these binders for stabilizing expansive soils.

The alkali-activated fly ash (AAFA) can be a promising alternative cementitious binder for stabilizing expansive soil. Fly ash, a prevalent by-product from thermal power plants, contains pozzolanic constituents such as silicates and aluminates. Typically, fly ash obtained from Indian power plants contains low calcium oxide content, classified as Class F fly ash (Chatterjee, 2011; Mypati & Saride, 2022; Syed et al., 2023). Given the minimal calcium oxide content in Class-F fly ash (FA), an alkaline environment is essential to induce the necessary pozzolanic reactions for activating the binding material. This alkaline environment within the fly ash is established by introducing a liquid alkaline activator (LAA) composed of a blend of liquid sodium silicate (Na_2SiO_3) solution, sodium hydroxide (NaOH), and ground granulated blast furnace slag (GGBS). The activation process of Class F FA includes the breakdown of aluminates and silicates within an alkaline environment, potentially resulting in the creation of a hydrated sodium–aluminum–silicate–hydrate (N–A–S–H) gel. Recently, numerous studies have explored the effects of AAFA binders in stabilizing expansive clays (Cristelo et al., 2011; Corrêa-Silva et al., 2018; Maio et al., 2017; Mypati & Saride, 2022; Rios et al., 2016; Syed et al., 2020, 2023; Wang et al., 2023). However, only a few studies have focused on the application of AAFA binders in deep soil mixing. Syed et al. (2020) investigated the influences of varying percentages of AAFA and various water-to-solid ratios at different curing periods on the swelling and compressibility characteristics of black cotton soil. Mypate & Saride (2022) studied the effect of binder ratio (LAA/FA) and curing period on the shear strength behaviour of stabilized expansive soil and found significant improvement in the unconfined compressive strength (UCS) of the treated expansive soils. Similar observations were also noted by Wang et al. (2023) based on the direct shear test and UCS test on expansive soils mixed with AAFA binders.

Given the limited research on the application of AAFA binders in deep soil mixing, this current study aims to investigate the impact of the length-to-diameter ratio (*l*/*d*) of deep mixing soil columns on the swelling potential and shear strength properties of expansive soil treated with AAFA binders through extensive laboratory investigation.

Materials

Expansive Clay (EC)

In this study, the EC used was obtained from the Ravulapalem, Konaseema district of Andhra Pradesh, India. The soil, classified as residual soil, was extracted from a depth of 1 m below the natural ground surface to prevent interference from vegetation or organic matter. The expansive clay was dried in an oven before being pulverized using a wooden mallet to remove lumps. The soil samples were then used in various experiments to determine the physical, chemical, index, and geotechnical properties. The specific physicochemical, index, and engineering characteristics of the expansive clay are presented in Table 1.

Table 1. Index and engineering properties of the expansive clay

Property	Expansive clay	Fly ash	Reference
Liquid limit, w_l (%)	74	21.1	ASTM D4318-17 (2018)
Plastic limit, w_p (%)	31.8	-	
Plasticity Index, PI (%)	42.2	-	
Sand (%)	11.3	1.8	
Silt (%)	36.2	68.4	
Clay (%)	52.5	29.8	
Free swell index	120	-	
Specific gravity, *G*	2.71	2.31	ASTM D854 (2016)
Optimum moisture content, OMC (%)	26.6	28.5	ASTM D698-12 (2021)
Maximum dry density, MDD (kN/m$_3$)	15.3	12.1	
USCS classification	CH	MH	
Unconfined compressive strength, UCS (kPa)	130	-	

Alkali Activated Fly Ash (AAFA)

Class F Fly ash (FA) was employed in this study. FA was collected from a thermal power plant in Vijayawada City (NTPC Vijayawada) of the Krishna district, Andhra Pradesh. The chemical composition of the fly ash was assessed using XRF analysis, and the results are presented in Table 2. In order to create the alkaline activating solution procedure suggested by Syed et al. (2020) was adopted in this study. Sodium hydroxide pellets with a purity of

99% and industrial-grade sodium silicate with 29.5% SiO_2 and 14.7% Na_2O by weight were employed.

Table 2. Chemical composition of FA

Oxides	Percentage (%)
SiO_2	59.55
Al_2O_3	29.52
Fe_2O_3	5.01
CaO	1.65
MgO	0.48
K_2O	2.14
TiO_2	1.65

Variables Studied

Laboratory experiments were carried out to investigate the performance of DSM columns in restraining swelling and enhancing load-bearing capabilities in EC. To evaluate this, the surface heave, radial heave, and load-bearing capacities of EC were measured under two conditions: without and with DSM columns. These measurements were obtained by conducting four distinct heave and load-bearing tests. The EC was compacted within California Bearing Ratio (CBR) molds.

- In all experimental tests, the compaction was done at 15% water content with a dry unit weight of 15 kN/m^3. The chosen conditions were specifically selected based on their suitability for compaction purposes. When EC is compacted below the OMC, it offers a higher degree of suction or capillary forces within the soil matrix. As a result, the clay particles tend to attract and hold water more aggressively, leading to enhanced swelling tendencies. Hence the water content was chosen below the OMC to understand and mitigate the effects of heave in EC.
- CBR molds used in this study have a diameter of 150 mm and a height of 175 mm, excluding the collar. The uniformity of the expanded clay beds' thickness was maintained at a constant value of 100 mm. The DSM column lengths (*l*) were maintained at 50 mm, 75 mm, and 100 mm. Nevertheless, the dimension of the DSM columns (*d*) remained

constant at 30 mm. To clarify, the DSM columns with aspect ratios of 1.66, 2.5, and 3.33 are employed.

Experimental Program

Preparation of EC and DSM Column

As depicted in Figure 1, the experimental set-up utilized a CBR mold, in which a layer of sand of 10 mm in thickness was placed at the base and subsequently levelled using a rammer. A circular box was formed by inserting a hollow cylindrical metal sheet with a diameter of 130 mm into small molds, ensuring a consistent distance of 10 mm from the walls of the tank. The gaps 10 mm in width were filled with sand in order to create a sand-packing system designed to facilitate the rapid attainment of heave of the expansive clay layer. More details on the compaction process are available from other sources (Raghuram et al., 2022).

A 30 mm diameter hollow metal sheet was inserted in the center of the EC so as to prepare the DSM column with varying lengths. The DSM columns with the above-mentioned *l*/*d* ratios were prepared with alkali-activated fly ash. Figure 2 shows the excavation made for the installation of DSM column.

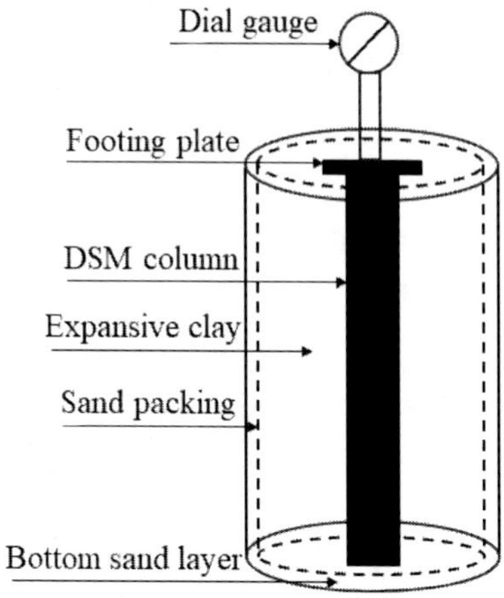

Figure 1. Experimental set up for compression tests on expansive clays.

Figure 2. Excavation made for DSM column.

Figure 3. Expansive clay bed stabilized with DSM column.

The FA was activated with sodium hydroxide, sodium silicates, and water. The water-to-solids (w/s) ratio and optimum percentage of alkali-activated fly ash (AAFA) were maintained at 0.4 and 10% (by weight of expansive clay) respectively as recommended by Syed et al. (2020). The total weight of AAFA turned out to be 150 kg/m^3 when expansive clay was compacted at 1500 Kg/m^3. The mass ratio of fly ash to sodium hydroxide to sodium silicate was considered to be 400:10.57:129.43 as suggested by Syed et al. (2020). The DSM columns were then prepared with varying lengths with the aforementioned AAFA (refer Figure 3).

Tests Performed

Surface and Radial Heave Tests

The unreinforced and reinforced EC were subjected to wetting through the continual addition of water to the soil. The measurement of surface and radial heave was conducted in a continuous manner using a dial gauge until the clay beds reached saturation or reached a state of equilibrium heave. Following the attainment of equilibrium heave, the water contents of specimens obtained from different depths of the clay layers were assessed in order to ascertain their entire saturation.

Compression Load Tests

A total of 4 plate load tests were performed to investigate the stress-settlement characteristics of three different soil conditions:

1. EC alone, and
2. EC reinforced with DSM column, where the bearing plate/footing plate is supported by both the clay and the DSM column (as illustrated in Figure 1).

A compressive load (N) was applied to the clay beds upon their saturation. For the load test on the composite ground (consisting of expansive clay and DSM column) a bearing plate with a diameter of 70 mm was employed. The description of the load tests was presented elsewhere (Raghuram et al., 2020).

Results and Discussion

Surface Heave

Figure 4 illustrates the heave rate observed in the unreinforced EC. Figure 4 illustrates the temporal changes in heave measurements, on the Y-axis (mm), during a given duration of time, on the X-axis (min.). The final equilibrium heave of 7.4 mm was seen in the EC within a duration of 2880 minutes, as evidenced by the experimental results. Additionally, it may be noted from

Figure 4 that the surface heave of the EC reduced drastically when the DSM columns were installed. This can be attributed to the installation of DSM columns that efficiently mix and interlock with the adjacent soil, thereby enhancing its shear strength and mitigating its vulnerability to heaving. This process mitigates the vertical expansion of the clay, as the columns function as barriers that hinder the upward movement of the soil.

As an illustration, the surface heave was reduced by 59.45% (7.4 mm to 3 mm), 68.92% (7.4 mm to 2.3 mm), and 75.67% (7.4 mm to 1.8 mm) respectively when DSM columns of *l/d* were varied as 1.66, 2.5, and 3.33. Additionally, as the l/d ratio of the DSM columns increases, the AAFA progressively penetrates the underlying expansive clay, thus augmenting the overall volume of the reinforced soil. The increase in the *l/d* ratio of the DSM column results in a larger zone of influence for the DSM columns, enhancing their ability to effectively mitigate the swelling forces exerted by the clay.

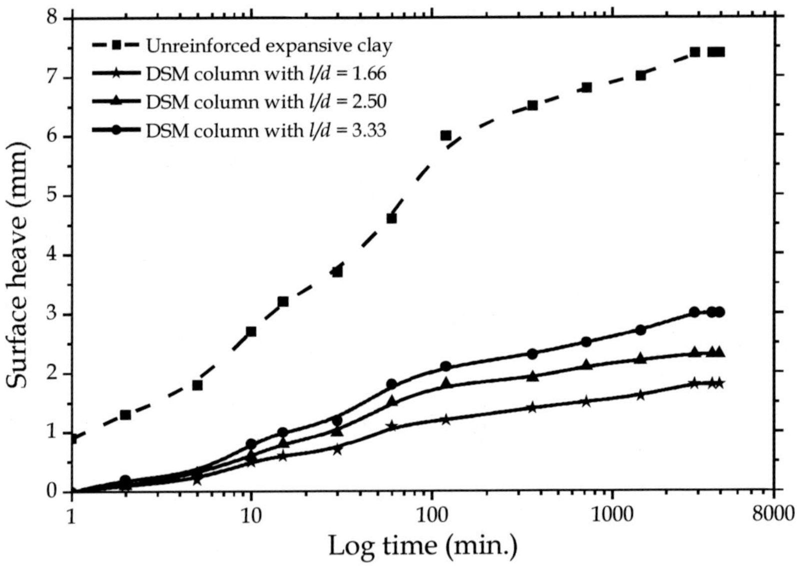

Figure 4. Effect of *l/d* ratio of DSM on the surface heave of the expansive clay.

Radial Heave

The effect of the *l/d* ratio of the DSM column on the radial heave is demonstrated in Figure 5. The final radial heave was measured at a distance of 15 mm, 30 mm, and 45 mm from the center of the DSM column at 2880 mins. It can be observed from Figure 5 that the radial heave of the EC increased as the radial distance from the center of the DSM column increased. This may be due to the influence of the treatment area of the DSM column. The DSM column largely enhances soil strength and stability in its immediate area. Hence, as the radial distance from the center of the DSM column increases, the EC material remains relatively untreated and susceptible to the effects of swelling and heaving pressures. As an example, for a given *l*/*d* ratio of 3.33, the surface heave increased from 1.8 to 6.3 when the radial distance from the center of the DSM column increased from 0 mm to 45 mm. The data presented in Figure 5 provides obvious evidence that the DSM column demonstrates an efficient reduction of surface heave at a radial distance of 1.5 times the diameter of the DSM column from its center.

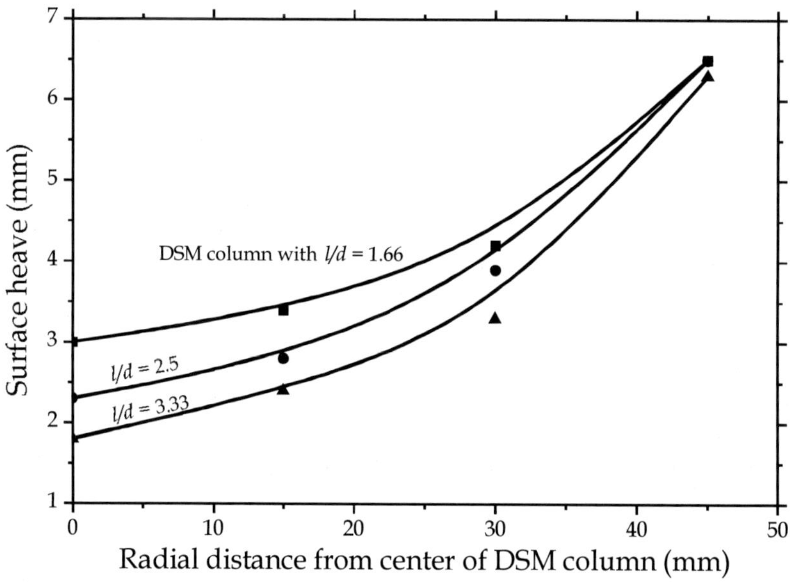

Figure 5. Influence of *l*/*d* ratio of DSM on the radial heave of the expansive clay.

Load Tests

The results shown in Figure 6 illustrate the influence of the *l*/*d* ratio of the DSM column on the load-carrying capacities of the EC. It can be noted from Figure 6 that the load-carrying capacities of the EC increased significantly when the clay was reinforced with a DSM column. It may to attributed to the inclusion of DSM columns significantly enhances the structural integrity of the clay material by augmenting its shear strength, mitigating its sensitivity to volumetric changes, and offering enhanced lateral reinforcement. Consequently, the soil exhibits enhanced load-bearing capacity, thereby minimizing the occurrence of excessive settlement or deformation. In addition, the use of DSM columns facilitates a more even distribution of applied stresses across the clay bed, hence mitigating the occurrence of localized stress concentrations and potential areas of failure.

For instance, for a given settlement of 20 mm, the load-carrying capacities were increased from 185 kPa to 902 kPa (by 4.87 times), 185 kPa to 1075 kPa (by 5.81 times), and 185 kPa to 1293 kPa (by 6.98 times) respectively when DSM columns of *l*/*d* were varied as 1.66, 2.5, and 3.33. This could be due to the fact that the increase in the *l*/*d* ratio of the DSM column offers increased lateral support to the soil, creating a stronger and more stable foundation. This, in turn, enables the expansive clay to withstand heavier loads without compromising its structural integrity. The DSM columns with a high *l*/*d* ratio also distribute the load more uniformly across the clay bed.

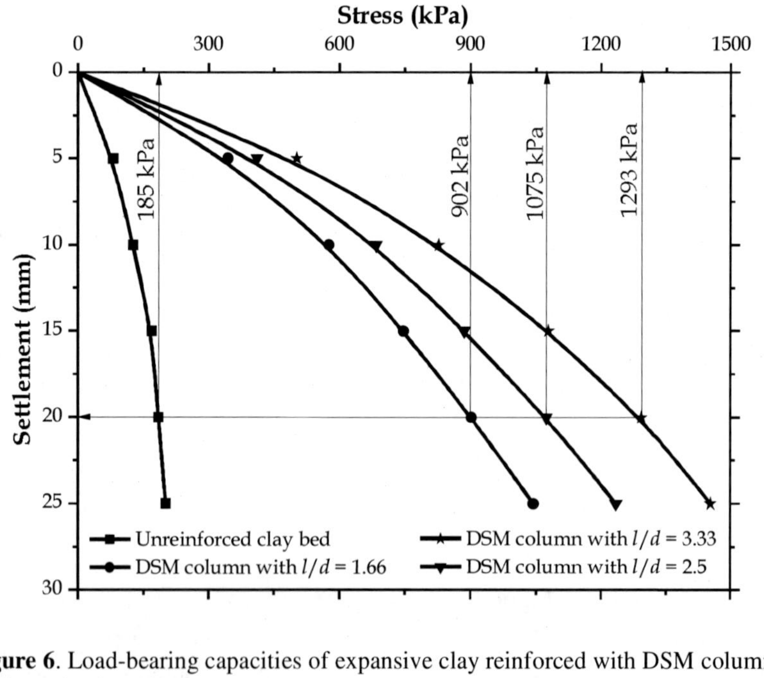

Figure 6. Load-bearing capacities of expansive clay reinforced with DSM columns with varying lengths.

Final Considerations

The objective of this research is to evaluate the effectiveness of alkali-activated fly ash (AAFA) in the stabilization of expansive clays (EC) using the in-situ deep soil mixing (DSM) method. The primary aim of this study is to address the swell-shrink behavior of expansive clay (EC) and achieve enhanced strength characteristics using environmentally friendly binding materials. For this purpose, low-calcium fly ash was activated with a mixture of sodium hydroxide and sodium silicate. Deep soil mixing columns were then prepared to stabilize the EC. The following are the primary conclusions that can be drawn from this study.

1. Increasing the *l*/*d* ratio of the DSM column resulted in a reduction in the surface heave of the EC. The surface heave was reduced from 7.4 mm to 1.8 mm when the DSM column of $l/d = 3.33$ was installed in the EC.
2. The surface heave increased with an increase in the radial distance from the center of the DSM column. The surface heave increased from 1.8 to 6.3 when the radial distance from the center of the DSM column increased from 0 mm to 45 mm.
3. Load-carrying capacities of EC increased with the increase in *l*/*d* ratios of the DSM column.
4. The load-carrying capacities of EC reinforced with DSM columns with *l*/*d* ratios of 1.66, 2.5, and 3.33 were 902 kPa, 1075 kPa, and 1293 kPa respectively.

The utilization of alkali-activated fly ash has demonstrated its efficacy as a binding agent in the process of deep soil mixing, leading to enhanced soil characteristics and mitigated surface heave. The results of this study provide evidence that alkali-activated fly ash has the potential to be utilized for construction projects focused on enhancing the performance of expansive clay soils.

References

Al-Tabbaa, A. & Evans, C. W. (1999). Laboratory-scale soil mixing of a contaminated site. *Proceedings of the Institution of Civil Engineers*, *3*(3), 119–134. https://doi.org/10.1680/gi.1999.030303.

ASTM. 2018. Standard test methods for liquid limit, plastic limit, and plasticity index of soils. ASTM D4318-17. West Conshohocken, PA: ASTM.

ASTM. 2016. Standard test methods for specific gravity of soil solids by water pycnometer. ASTM D854. West Conshohocken, PA: ASTM.

ASTM. 2021. Standard test methods for laboratory compaction characteristics of soil using standard effort (12 400 ft-lbf/ft^3 (600 kN-m/m^3)). ASTM D698-12. West Conshohocken, PA: ASTM.

Barman, D. & Dash, S. K. (2022). Stabilization of expansive soils using chemical additives: A review. *Journal of Rock Mechanics and Geotechnical Engineering*, *14*(4), 1319–1342. https://doi.org/10.1016/j.jrmge.2022.02.011.

Bhadriraju, V., Puppala, A. J., Madhyannapu, R. S. & Williammee, R. (2008). Laboratory procedure to obtain Well-Mixed soil binder samples of medium stiff to stiff expansive clayey soil for deep soil mixing simulation. *Geotechnical Testing Journal*, *31*(3), 100936. https://doi.org/10.1520/gtj100936.

Considine, M. (1984). Soils shrink, trees drink, and houses crack. *ECOS Magazine*, *41*, 13-15.

Chatterjee, A. (2011). Indian Fly ashes: Their characteristics and potential for mechanochemical activation for enhanced usability. *Journal of Materials in Civil Engineering*, *23*(6), 783–788. https://doi.org/10.1061/(asce)mt.1943-5533.0000279.

Corrêa-Silva, M., Araújo, N. M. F., Cristelo, N., Miranda, T. F. S., Gomes, A. T. & Coelho, J. (2018). Improvement of a clayey soil with alkali activated low-calcium fly ash for transport infrastructures applications. *Road Materials and Pavement Design*, *20*(8), 1912–1926. https://doi.org/10.1080/14680629.2018.1473286.

Cristelo, N., Glendinning, S. G., Fernandes, L. & Pinto, A. T. (2012). Effect of calcium content on soil stabilisation with alkaline activation. *Construction and Building Materials*, *29*, 167–174. https://doi.org/10.1016/j.conbuildmat.2011.10.049.

Dang, L. C., Hasan, H., Fatahi, B. & Khabbaz, H. (2015). Influence of bagasse ash and hydrated lime on strength and mechanical behaviour of stabilised expansive soil. *GeoQuebec 2015*.

Eujine, G. N., Chandrakaran, S. & Sankar, N. (2017). Accelerated subgrade stabilization using enzymatic lime technique. *Journal of Materials in Civil Engineering*, *29*(9). https://doi.org/10.1061/(asce)mt.1943-5533.0001923.

FHWA. 2013. Federal highway administration design manual: Deep mixing for embankment and foundation support. FHWA-HRT-13-046. McLean, VA: FHWA.

Holtz, R. D., Kovacs, W. D. & Sheahan, T. C. (2011). *An introduction to geotechnical engineering*. (2nd ed.) Prentice Hall.

Jain, A., Choudhary, A. K. & Jha, J. N. (2019). Influence of rice husk ash on the swelling and strength characteristics of expansive soil. *Geotechnical and Geological Engineering*, *38*(2), 2293–2302. https://doi.org/10.1007/s10706-019-01087-6.

Jones, L. D. & Jefferson, I. (2012). *Expansive soils*. (J. Burland (ed.); 1st ed.). ICE Publishing.

Li, J., Cameron, D. A. & Ren, G. (2014). Case study and back analysis of a residential building damaged by expansive soils. *Computers and Geotechnics*, *56*, 89–99. https://doi.org/10.1016/j.compgeo.2013.11.005.

Lorenzo, G. A. & Bergado, D. T. (2006). Fundamental characteristics of cement-admixed clay in deep mixing. *Journal of Materials in Civil Engineering*, *18*(2), 161–174. https://doi.org/10.1061/(asce)0899-1561(2006)18:2(161).

Madhyannapu, R. S. & Puppala, A. J. (2014). Design and construction guidelines for deep soil mixing to stabilize expansive soils. *Journal of Geotechnical and Geoenvironmental Engineering*, *140*(9). https://doi.org/10.1061/(asce)gt.1943-5606.0001149.

Miao, S., Shen, Z., Wang, X., Luo, F., Huang, X. & Wei, C. (2017). Stabilization of highly expansive black cotton soils by means of geopolymerization. *Journal of Materials in Civil Engineering*, *29*(10). https://doi.org/10.1061/(asce)mt.1943-5533.0002023.

Mir, B. A. (2002). *The effect of fly ash on the engineering properties of black cotton soils* (master's thesis). Indian Institute of Science Bangalore.

Mypati, V. N. K. & Saride, S. (2022). Feasibility of alkali-activated low-calcium fly ash as a binder for deep soil mixing. *Journal of Materials in Civil Engineering*, *34*(1). https://doi.org/10.1061/(asce)mt.1943-5533.0004047.

Okagbue, C. O. & Onyeobi, T. (1999). Potential of marble dust to stabilise red tropical soils for road construction. *Engineering Geology*, *53*(3–4), 371–380. https://doi.org/10.1016/s0013-7952(99)00036-8.

Petry T. M. & Armstrong, J. C. (1989). *Stabilization of expansive clay soils*. Transportation Research Record 1219, Transportation Research Board.

Phanikumar, B. R., Raghuram, A. S. S. & Sriramarao, A. (2020). Improving expansive clay beds with granular pile anchors (GPAs) and geogrid-encased GPAs. *Proceedings of the Institution of Civil Engineers*, *173*(4), 237–248. https://doi.org/10.1680/jgrim.18.00086.

Raghuram, A. S. S., Phanikumar, B. R., Purnanandam, K. & Rao, A. J. (2022). Heave and pullout studies of geogrid-encased granular pile-anchors in expansive clay beds. *Proceedings of the Institution of Civil Engineers*, *175*(4), 236–246. https://doi.org/10.1680/jgrim.19.00081.

Raja, P. S. K. & Thyagaraj, T. (2020). Sulfate effects on sulfate-resistant cement–treated expansive soil. *Bulletin of Engineering Geology and the Environment*, *79*(5), 2367–2380. https://doi.org/10.1007/s10064-019-01714-9.

RíOs, S., Cristelo, N., Da Fonseca, A. V. & Ferreira, C. (2016). Structural performance of alkali-activated soil ash versus soil cement. *Journal of Materials in Civil Engineering*, *28*(2). https://doi.org/10.1061/(asce)mt.1943-5533.0001398.

Sahoo, J. P. & Pradhan, P. K. (2010). Effect of lime stabilized soil cushion on strength behaviour of expansive soil. *Geotechnical and Geological Engineering*, *28*(6), 889–897. https://doi.org/10.1007/s10706-010-9332-6.

Saride, S., Puppala, A. J. & Chikyala, S. R. (2013). Swell-shrink and strength behaviors of lime and cement stabilized expansive organic clays. *Applied Clay Science*, *85*, 39–45. https://doi.org/10.1016/j.clay.2013.09.008.

Sherwood, P. T. (1994). *Soil stabilization with cement and lime*. Her Majesty Stationary Office.

Sidiq, A., Robert, D., O'Donnell, B., Setunge, S., Swarup, A. & Tostvrsnik, F. (2023). Investigation of Enzyme–Based soil stabilization in field application. *Journal of Materials in Civil Engineering*, *35*(5). https://doi.org/10.1061/(asce)mt.1943-5533.0004742.

Syed, M., GuhaRay, A. & Kar, A. (2020). Stabilization of expansive clayey soil with alkali activated binders. *Geotechnical and Geological Engineering*, *38*(6), 6657–6677. https://doi.org/10.1007/s10706-020-01461-9.

Syed, M., GuhaRay, A. & Raju, S. (2023). Subgrade strength performance behavior of Alkali-Activated binder and cement stabilized expansive soil: a semifield study. *Journal of Materials in Civil Engineering*, *35*(10). https://doi.org/10.1061/jmcee7.mteng-15580.

Taha, M. M. M., Feng, C. & Ahmed, S. (2021). Modification of mechanical properties of expansive soil from north China by using rice husk ash. *Materials*, *14*(11), 2789. https://doi.org/10.3390/ma14112789.

Tanyıldızı, M., Uz, V. E. & Gökalp, İ. (2023). Utilization of waste materials in the stabilization of expansive pavement subgrade: An extensive review. *Construction and Building Materials*, *398*, 132435. https://doi.org/10.1016/j.conbuildmat.2023.132435.

Thyagaraj, T., Rao, S. M., Suresh, P. S. & Salini, U. (2012). Laboratory studies on stabilization of an expansive soil by lime precipitation technique. *Journal of Materials in Civil Engineering*, *24*(8), 1067–1075. https://doi.org/10.1061/(asce)mt.1943-5533.0000483.

Wang, H., Liu, T., Yan, C. & Wang, J. (2023). Expansive soil stabilization using alkali-activated fly ash. *Processes*, *11*(5), 1550. https://doi.org/10.3390/pr11051550.

Xu, B. & Yi, Y. (2021). Soft clay stabilization using three industry byproducts. *Journal of Materials in Civil Engineering*, *33*(5). https://doi.org/10.1061/(asce)mt.1943-5533.0003710.

Zada, U., Jamal, A., Iqbal, M., Eldin, S. M., Almoshaogeh, M., Bekkouche, S. R. & Almuaythir, S. (2023). Recent advances in expansive soil stabilization using admixtures: current challenges and opportunities. *Case Studies in Construction Materials*, *18*, e01985. https://doi.org/10.1016/j.cscm.2023.e01985.

Chapter 10

Stabilization of Iron Ore Tailings with Alkaline Cements

Suéllen Tonatto Ferrazzo[1,*]
William Mateus Kubiaki Levandoski[2]
Maria Alice Piovesan[3]
and Eduardo Pavan Korf[2]
[1]Graduate Program in Civil Engineering, Universidade Federal do Rio Grande do Sul, Porto Alegre, RS, Brazil
[2]Graduate Program in Environmental Science and Technology, Universidade Federal da Fronteira Sul, Erechim, RS, Brazil
[3]Environmental and Sanitary Engineering, Universidade Federal da Fronteira Sul, Erechim, RS, Brazil

Abstract

Mining is an important sector of the world economy, with a focus on iron ore in Brazil. With the accelerated growth of the mining industry comes the problem of the large amounts of tailings generated during the processing of ores. For example, 1 ton of iron processed generates about 400 kg of tailings, which is generally disposed of in dams. In addition, there are the environmental impacts and risks associated with traditional waste disposal, which can lead to tragedies such as those that occurred in Mariana and Brumadinho, state of Minas Gerais, Brazil. In this context, the stabilization/solidification of iron ore tailings (IOT) with alkaline cements, considered as more "sustainable" alternatives to Portland cement, has been studied. In this sense, the contribution of this chapter is to provide an overview of the research on IOT stabilization with alkaline

* Corresponding Author's Email: suellenferrazzo@hotmail.com.

In: Environmental Geotechnics
Editors: E. Pavan Korf, P. Domingos Marques Prietto, C. Müller et al.
ISBN: 979-8-89113-651-9

cements in a geotechnical context and to discuss the main results, knowledge gaps and trends. As a starting point, metakaolin and alkaline activators, both of commercial origin, have been used in the production of cements by the “one-part” or “two-part” methods for application in IOT. Precursors, alkaline activators and calcium sources from waste are a future trend, such as alternative metakaolin, rice husk ash, sugarcane bagasse ash, alternative sodium silicate, carbide lime, and eggshell lime. The mechanical behavior of IOT-cement mixtures has been extensively studied and some studies have also evaluated mineralogy and morphology. Recently, the environmental impact of stabilizing mine tailings with various alkali-activated cements has been researched using life cycle assessments. In general, stabilization of IOT with alkaline cements resulted in satisfactory mechanical and environmental performance. However, the leaching of pollutants and the economic and social aspects of sustainability remain unexplored.

Keywords: mine tailings, alkali-activated cements, mechanical behavior, life cycle assessment

Iron Mining Industry

World iron ore reserves are estimated at 170 billion tons, with the largest deposits in Australia, Brazil, China, Ukraine and Russia. In Brazil, the most important with iron ore reserves are Minas Gerais, Pará, and Mato Grosso do Sul (De Jesus & Joaquim, 2020). Global iron ore production increased from 274 million tons in 1950 to 1.5 billion tons in 2005 (Yellishetty et al., 2010) and 2.4 billion tons in 2017 (De Jesus & Joaquim, 2020). Brazil produced about 450 million tons of iron ore in 2017 (De Jesus & Joaquim, 2020).

Australia, Brazil, and China are the main exporters of iron ore (Yellishetty et al., 2010). In 2017, Brazil exported about 350 million tons of iron ore (De Jesus & Joaquim, 2020), representing approximately 83% of Brazil´s total ore exports (MME, 2019). Japan is the largest importer of iron ore in the world, followed by China and Europe (Yellishetty et al., 2010).

The main product produced from iron ore is steel. This metallic alloy is used to produce a wide range of materials in various industrial sectors. It is estimated that 1.3 tons of iron ore are required to produce 1.0 ton of steel (Tuck & Ghalayini, 2016). China, the world’s largest steel producer, produced about 356 million tons of steel in 2017 (Wang et al., 2020). In the same year, Brazil produced around 35 million tons of steel (MME, 2019).

Due to the low iron content of the ore (31-45 wt% Fe, iron quadrangle), processing operations are carried out. The iron ore destined for the steel industry undergoes fragmentation, size separation, and concentration (Dauce et al., 2019). According to Dauce et al. (2019), the processing of 1 ton of iron ore generates about 400 kg of waste.

Iron Ore Tailings

It is estimated that 1.4 billion tons of iron ore tailings are generated annually worldwide, mainly in Australia, Brazil, China and India (Carmignano et al., 2021). In Brazil, about 560 tons of IOT are generated per year in the state of Minas Gerais alone (Carmignano et al., 2021). This amount is distributed in 672 tailings dams, 340 of which are located in Minas Gerais (Carmignano et al., 2021; Fontes et al., 2019). The tailings dams occupy large areas and have significant visual impacts throughout their useful life (Fontes et al., 2019), causing environmental impacts such as deforestation (during construction), pollution of soil and surface water and underground (during operation and after mine closure) due to leaching of pollutants (Maruthupandian et al., 2021), etc. In case of rupture, they cause environmental, economic, and social impacts of large scale and complex measurement (Becker et al., 2019; Davila et al., 2020; Edraki et al., 2014; Kossoff et al., 2014). The risks associated with these geotechnical structures and the exploitation of lower-quality iron ore deposits (i.e., lower iron content) have led to a significant increase in the number of tailings-related accidents (Fontes et al., 2019).

Between 2015 and 2019, Brazil recorded two major mining tailings dam disasters, Mariana (2015) and Brumadinho (2019), both in Minas Gerais (Freitas & Da Silva, 2019). The Fundão tailings dam (Mariana) collapse is considered the largest mining disaster in the world in terms of volume of tailings released and geographic extent affected (Carmo et al., 2017). The rupture resulted in 19 deaths (Freitas & Da Silva, 2019), released about 44 million m^3 of IOT, and the damage caused by pollutants extended over 680 km of watercourses from the Doce River to the Atlantic Ocean, reaching 39 municipalities in the states of Minas Gerais and Espírito Santo (Carmo et al., 2017). The Brumadinho disaster caused 300 victims (including fatalities and missing persons) and released 13 million m^3 of IOT, affecting 18 municipalities over a length of 250 km (Freitas & Da Silva, 2019).

The increase in accidents in mine dams, combined with the need to build new geotechnical structures (the existing ones are reaching their full capacity) is leading to a scientific and socioeconomic appeal for the correct and safe management of iron mine tailings (Davila et al., 2020; Fontes et al., 2019). Therefore, there is a growing interest in techniques that allow the reuse or recycling of mining tailings, given the potential benefits of reducing final disposal, minimizing environmental impacts, and adding value to these wastes (Araujo et al., 2022).

Chemical Stabilization of Iron Ore Tailings

To improve the mechanical properties of IOT and reduce the risks associated with traditional tailings disposal, chemical stabilization of IOT by addition of binders was investigated. Stabilization/solidification of mining wastes can be considered as an effective method to treat wastes, minimizing the risks associated with metal leaching (Maruthupandian et al., 2021; Pan et al., 2022). There are numerous research papers in the literature on IOT stabilized with different binders, such as fly ash (Manjunatha & Sunil, 2013), commercial lime (Machado, 2022; Manjunatha & Sunil, 2013), bio-cementation (Buikema, 2015), polymer (Carneiro, 2020), bentonite (Barati et al., 2020; Machado, 2022), steel aggregates, mining wastes (Machado, 2022), and mainly with Portland cement (Barati et al., 2020; Carvalho et al., 2023; Chaves, 2023; Consoli et al., 2022; Gentil, 2020; Machado, 2022; Mafessoli, 2022; Manjunatha & Sunil, 2013).

Studies show that the incorporation of small amounts (e.g., 1-5%) to 15% Portland cement in the IOT positively and significantly affects the initial stiffness (Carvalho et al., 2023; Chaves, 2023; Consoli et al., 2022; Mafessoli, 2022; Mafessoli et al., 2023), the unconfined compressive strength (Barati et al., 2020; Carvalho et al., 2023; Chaves, 2023; Consoli et al., 2022; Machado, 2022; Mafessoli, 2022; Mafessoli et al., 2023) and the split tensile strength (Chaves, 2023; Mafessoli, 2022) of the cemented mixture. The addition of 3% Portland cement enhances the strength and stiffness properties of IOT for low confining stress (300 kPa) (Bruschi et al., 2023). Furthermore, the unconfined compressive strength of IOT-cement increased with the increase of binder content in the mixture (Barati et al., 2020; Chaves, 2023; Machado, 2022; Mafessoli, 2022) and curing time (Barati et al., 2020; Chaves, 2023; Mafessoli, 2022), indicating the great effectiveness of Portland cement in stabilizing mining tailings (Mafessoli, 2022). The addition of traditional cement to IOT is also an interesting alternative for use in the dry stacking method, as well as for the safe disposal of mining tailings (Bruschi et al., 2023;

Carvalho et al., 2023; Consoli et al., 2022). IOT stabilized with Portland cement also meets the standards required for the use of the cemented material in road base and subbase layers (Barati et al., 2020).

However, the Portland cement production chain is associated with numerous negative impacts, such as the depletion of natural resources, energy consumption, and the generation of CO_2 emissions (Garcia-Lodeiro et al., 2015; Moretti et al., 2016). The production of 2 tons of cement requires more than 3 tons of raw material (about 70% limestone) (Garcia-Lodeiro et al., 2015), which is calcined at high temperatures (1450-1550 °C) (Garcia-Lodeiro et al., 2015; Huntzinger & Eatmon, 2009). For every ton of Portland cement produced, 1 ton of CO_2 is generated, representing 8% of global greenhouse gas emissions, making manufacturing one of the most significant activities contributing to global warming (Andrew, 2019).

In this context, research has been conducted to develop alternative binders that have a lower environmental impact and perform comparably to conventional cement. Alkaline cements (commonly known) represent a more "sustainable" category of cements and are produced through the alkaline activation process of an aluminosilicate source (precursor), such as fly ash, metakaolin, blast furnace slag, and numerous industrial wastes (Garcia-Lodeiro et al., 2015; Provis & Van Deventer, 2014) (see Chapter 4).

Iron Ore Tailings Stabilized with Alkaline Cements

In recent years, several alkali-activated binders (AABs) and geopolymers have been developed and used to stabilize IOT, focusing on mechanical performance and microstructure, as well as for potential use in dry stacking of mining tailings. Figure 1 provides an overview of the precursor materials, calcium sources, and activators used in the production of alkaline cements, as well as evaluations performed on IOT-cement mixtures. The studies on this topic are then presented, followed by a discussion of the main findings and knowledge gaps.

The first study on IOT stabilization with alkaline cements refers to the use of AAB based on metakaolin, a product resulting from the calcination of kaolin and used as a traditional precursor of the low-calcium system (see Chapter 4). To stabilize IOT, Fracaro (2022) combined metakaolin with sodium silicate and sodium hydroxide in the development of an AAB using the "two-part" method (i.e., precursor in solid state and alkaline activator in liquid state). When identical percentages of AAB and high initial strength

Portland cement (type III) were compared on the wastes, it was found that the metakaolin-based binder provided higher unconfined compressive strength with longer curing (28 and 60 days) as the strength increased with time. Cementation of the IOT-AAB mixture was demonstrated by the formation of sodium aluminosilicate hydrate gel (N-A-S-H) as a geopolymerization product, which is different from the calcium silicate hydrate gel (C-S-H) formed when IOT is stabilized with traditional cement.

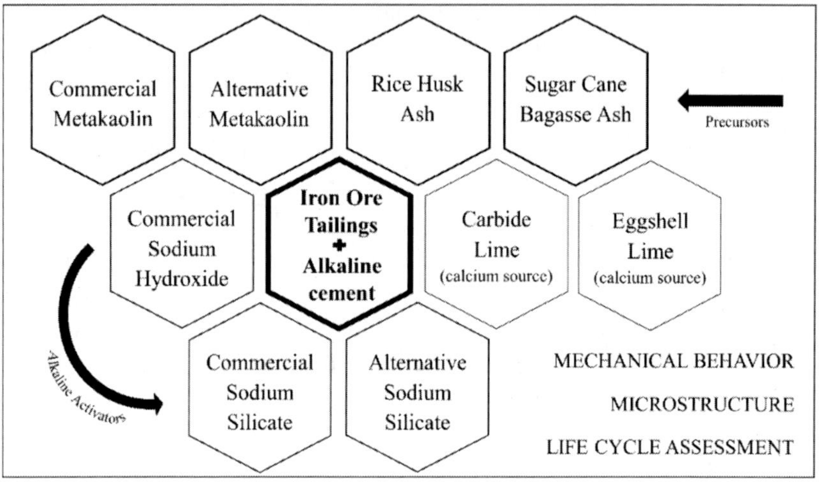

Figure 1. Overview of research on IOT stabilization with alkaline cements.

In a similar study, Jaskulski (2022) developed a geopolymer using the "one-part" method (i.e., precursor and alkaline activator in solid state) to stabilize IOT. This geopolymer is prepared from metakaolin activated by sodium hydroxide and sodium silicate. The author found that the mechanical behavior of IOT improved with the addition of 5% geopolymer and competitive strengths were obtained compared to those of high initial strength Portland cement (type III), although somewhat lower. It was found that the increase in cementation resulted in an increase in the initial stiffness and brittleness of the cemented IOT mixtures. In addition, scanning electron microscopy (SEM) analyzes with energy dispersive X-ray spectroscopy (EDS) indicated the presence of an N-A-S-H type cementation gel in the IOT-geopolymer matrix.

Farenzena et al. (2024) evaluated the stabilization of IOT with an AAB produced from alternative metakaolin and sodium silicate (i.e., a precursor and alkaline activator from mining waste) using the "one-part" method. The author found that increasing the ratio of precursor/activator reduced the unconfined compressive strength of IOT-AAB mixtures after 7 days of curing, possibly related to the formation of silica-rich cement gels and retarding the formation of zeolites. In addition, an increase in undrained shear strength and initial shear modulus were observed, as well as a decrease in the tendency to compress IOT-AAB mixtures due to the increase in binder content (0-5%). Microstructural analysis by SEM and X-ray diffraction (XRD) showed the presence of N-A-S-H gel forming bonds between the tailings particles, justifying the observed satisfactory mechanical behavior.

In the work of Fracaro (2022), Jaskulski (2022), and Farenzena et al. (2024), it is important to highlight the practicality and potential for practicality of alkaline cements developed at room temperature (23±2°C) which can help reduce the costs and impacts associated with energy consumption. The "onde-part" method used in the research of Jaskulski (2022) and Farenzena et al. (2024) allows the mixing of precursor and alkaline activator, both in solid state, followed by the addition of water to achieve the binding properties, similar to the use of Portland cement (Ye et al., 2016). This method does not require machinery and labor to prepare the alkaline activator solution on site minimizing cost and time, which is very attractive from the perspective of large-scale application. In addition, metakaolin produced from mining waste (Farenzena et al., 2024) can be a substitute for commercial metakaolin in the production of new cements. The use of alkaline activator from waste (Farenzena et al., 2024) is very promising to replace traditional alkaline activators, as the latter may involve high environmental impacts and production costs, such as commercial sodium hydroxide and sodium silicate (Zahmak et al., 2021). Finally, the aforementioned mechanical behavior studies have shown that AAB and metakaolin-based geopolymers are competitive alternatives to Portland cement for stabilizing IOT and to the traditional method of disposing of mining tailings.

It is worth highlighting the following research gaps identified in the stabilization of IOT with metakaolin-based cements. It is also necessary to investigate the use of calcium sources in these alkali-activated systems, especially those derived from wastes (e.g., carbide lime and eggshell lime), which may affect the consumption of alkaline activator and metakaolin and the behavior of the cemented IOT. It is also suggested that leaching tests be performed to evaluate the encapsulation of metals present in the waste to ensure a cemented material that does not pose environmental risks. In addition, it is important to perform life cycle analyzes to determine the sustainability of these alkaline cements used in IOT compared to the use of conventional binders such as Portland cement.

In the context of IOT treatment, there is also research focused on the use of AAB produced from industrial, agro-industrial, or mining wastes. Servi et al. (2022) studied the mechanical behavior of IOT stabilized with alkali-activated cement made from rice husk ash and carbide lime. Based on the results on the strength and durability (wetting-drying cycles) of the cemented tailings, the authors noted a better performance of the alternative binder compared to Portland cement, as well as the potential use in an alternative final disposal method "dry stacking". The formation of aluminosilicate hydrates from calcium and/or sodium (C, N)-A-S-H gel was also observed after 28 days of curing. The binder content in the IOT was found to have the greatest influence on the mechanical behavior, while the initial moisture content and curing time were of secondary importance. Moreover, the mechanical behavior results were correlated with the porosity/volumetric binder content index (η/B_{iv}), previously proposed by (Consoli, Bittar Marin, et al., 2019; Consoli, Leon, et al., 2019). Servi et al. (2022) found that the η/B_{iv} index is excellent at predicting the mechanical performance of cemented IOT by providing dosage curves for different binders (alternative and traditional cements), curing times, and mechanical responses.

In the study by Servi et al. (2022), an AAB was prepared from rice husk ash and carbide lime: rich sources of aluminosilicate and calcium, respectively, that have no added value and are generally disposed in industrial landfills (Pelisser et al., 2023; Reis et al., 2022). Reusing or recycling waste to develop a new cement can provide income to the industy and reduce the costs and impacts associated with final disposal in landfills, contributing to the circular economy (Ferrazzo et al., 2023). Moreover, from a practical point of view, the prediction of the mechanical behavior of IOT-binder by the η/B_{iv} index allows a reduction in the number of experiments and consequently in the consumption of materials and time. It is expected that the research of Servi et al. (2022) can further advance the chemical confirmation of the type of cement gel formed in the alkali-activated system using Fourier transform infrared spectroscopy (FTIR). In addition, pollutant leaching tests and life cycle analysis are important to provide answers on the environmental impact and sustainability of AAB-stabilized IOT based on agro-industrial and industrial wastes.

Table 1. Studies on iron ore tailings stabilized with alkaline cements

Precursor/calcium source	Alkaline activator	Evaluations	Highlights	Reference
Metakaolin	Sodium silicate + sodium hydroxide	Unconfined compressive strength Shear strength Stress-strain behavior Microstructure (SEM/EDS)	Improvement of mechanical properties Unconfined compressive strength values higher than Portland cement in 28 and 60 days N-A-S-H gel formation	Fracaro (2022)
Metakaolin	Sodium silicate + sodium hydroxide	Unconfined compressive strength Stress-strain behavior Microstructure (SEM/EDS)	Improvement of mechanical properties Unconfined compressive strength values slightly lower than Portland cement N-A-S-H gel formation	Jaskulski (2022)
Rice husk ash + carbide lime	Sodium silicate + sodium hydroxide	Unconfined compressive strength Split tensile strength Durability η/Biv index Microstructure (SEM/EDS)	Improvement of mechanical behavior Sustainable alternative for soil stabilization Mechanical behavior prediction Potential use in the dry stacking method	Servi et al. (2022)
Alternative metakaolin	Alternative sodium silicate	Unconfined compressive strength Simple shear Initial stiffness Microstructure (SEM-XRD)	Improvement of mechanical properties Increased undrained shear strength and reduced compression tendency N-A-S-H gel formation	Farenzena et al. (2024)
Sugarcane bagasse ash + hydrated eggshell lime	Sodium hydroxide	Unconfined compressive strength Split tensile strength Initial stiffness Microstructure (XRD/FTIR/SEM)	Satisfactory strength and stiffness Mechanical behavior prediction (η/Biv) N-A-S-H gel formation	Levandoski et al. (2023)
Alternative metakaolin x Commercial metakaolin	Alternative sodium silicate x Commercial sodium silicate	Unconfined compressive strength Life cycle assessment (environmental pillar)	Alternative AAB with better mechanical-environmental performance Potential use in the dry stacking method	Saldanha et al. (2023)

In the context of using residues in alkaline cements for IOT stabilization, sugarcane bagasse ash and eggshells are agro-industrial wastes with no added value, as they are generally disposed in industrial and landfills. However, these residues are rich sources of aluminosilicates and calcium carbonate that can be used in alkali-activated systems (Ferrazzo et al., 2023). In this sense, Levandoski et al. (2023) used an alkali-activated binder of sugarcane bagasse ash and hydrated eggshell lime to stabilize/solidify IOT. In this study, it was found that the cemented mining tailings had satisfactory strength and initial stiffness, reaching 6.59 MPa after 28 days at 40°C. The authors noted the significant positive influence of higher binder content (15% to 25%), dry unit weight and curing temperature. A good ability to predict the mechanical behavior of IOT-binder mixtures by the η/B_{iv} index and the formation of N-A-S-H gel was also found. With respect to the existing literature, the study by Levandoski et al. (2023) is a step forward in exploring the effect of curing temperature on the treatment of IOT with alkaline cement, as weel as on the microstructure and mechanical behavior of the material. However, it is still necessary to investigate the durability, leaching behavior of the IOT-binder, and sustainability of the stabilization/solidification process.

Recently, the environmental impact of IOT stabilization has become a topic of interest. Saldanha et al. (2023) conducted a life cycle assessment for IOT stabilized with alkali-activated cement made from metakaolin (produced from mining waste) and sodium silicate powder from flotation process waste (Cement 1). In comparison, the use of commercial metakaolin and sodium silicate produced by the hydrothermal spray drying process (Cement 2) was evaluated. Cement 1 was found to be more sustainable due to lower consumption of energy and natural resources, focusing on the use of mining waste in the alternative production of metakaolin. In addition, strength tests showed that Cement 1 is equivalent to commercial product-based cements. The authors recommend future research with the following objectives: reduction of sodium hydroxide content; impact of industrial scale sodium silicate production; better understanding of mechanical behavior and applications of cemented waste in practice.

The study by Saldanha et al. (2023) makes progress in evaluating the environmental impacts associated with the mechanical behavior of IOT stabilized with alkali-activated cements, examining commercial precursors and alkaline activators, as well as other waste-derived materials. The authors suggested different dosages for the IOT-cement mixture based on the correlation of the unconfined compressive strength results and the η/B_{iv} index. However, in addition to strength, other mechanical properties such as stiffness and durability can also be considered. In this sense, the responses to metal leaching can be related to the mechanical performance considered satisfactory to investigate whether the cemented material, which is technically feasible, does not pose environmental risks for applications in geotechnical construction. In addition, the economic and social pillars still need to be evaluated to determine the sustainability index of different IOT stabilization strategies using alkaline cements and to verify whether these binders are more sustainable compared to the use of Portland cement.

Table 1 provides a summary of the work discussed to date, which allow us to identify not only a number of promising results, but also important trends. The initial studies on stabilizing IOT with alkaline cements used precursors (metakaolin) and commercially available alkaline activators (sodium silicate and sodium hydroxide), which is interesting as a starting point since these are materials with a chemical composition and other standardized properties that are taken for granted by the industry. This allows better control of the variables involved in the production of the new cement and its application.

Subsequently, there is interest in the use of precursors and alkaline activators from wastes. This lead to the inclusion of new factors in the system, as wastes are variable in their properties and composition, making the optimization of binders and stabilization of mining tailings more challenging. As a future prospect, the development of many new cements to stabilize IOT is expected, considering the availability of a wide range of waste sources for aluminosilicates and calcium, as well as highly alkaline wastewaters for use as alkaline activators (see Chapter 4).

Regarding the evaluation of IOT-cement mixtures (Table 1), all the works dealt with studies of mechanical behavior (macro) and microstructure with analyzes of morphology and/or mineralogy. Recently, the environmental impact of IOT stabilization strategies with alkaline cements has been evaluated through life cycle assessments, which is an extremely relevant initiative to consider sustainability related to mechanical performance. However, the use of mining tailings and other residues requires leaching tests (see Chapter 11). Therefore, future research on this topic is expected to examine the encapsulation/inerting of contaminants and the economic and social pillars of IOT stabilization strategies.

Final Considerations

This chapter provided an overview of IOT stabilization with alkaline cements in a geotechnical context, discussing highlights, knowledge gaps and trends. In terms of mechanical performance, alkaline cements produced with precursors and activators of commercial origin or from waste have emerged as competitive alternatives to Portland cement for IOT stabilization. Alkaline cements produced from waste also stand out as having a lower environmental impact.

As mining tailings and other residues are involved, it is anticipated that future research will investigate the encapsulation of contaminants in cemented IOT, as well as the economic and social pillars of sustainability of stabilization strategies. Finally, it is expected that the information discussed in this chapter will serve as a reference for decision makers in academia, industry and government on issues related to the sustainability, management, and treatment of mining tailings with an emphasis on waste valorization.

References

Andrew, R. M. (2019). Global CO_2 emissions from cement production, 1928–2018. *Earth System Science Data*, 10, 2213–2239. https://doi.org/10.5194/essd-2019-152.

Araujo, F. S. M., Taborda-Llano, I., Nunes, E. B. & Santos, R. M. (2022). Recycling and reuse of mine tailings: a review of advancements and their implications. *Geosciences*, *12*(9), 319. https://doi.org/10.3390/geosciences12090319.

Barati, S., Tabatabaie Shourijeh, P., Samani, N. & Asadi, S. (2020). Stabilization of iron ore tailings with cement and bentonite: A case study on Golgohar mine. *Bulletin of Engineering Geology and the Environment*, *79*(8), 4151–4166. https://doi.org/10.1007/s10064-020-01843-6.

Becker, F., Eser, R. A., Hoelzmann, P. & Schütt, B. (2019). The environmental impact of ancient iron mining and smelting on Elba Island, Italy – A geochemical soil survey of the Magazzini site. *Journal of Geochemical Exploration*, *205*, 106307. https://doi.org/10.1016/j.gexplo.2019.04.009.

Bruschi, G. J., Pereira dos Santos, C., Scheuermann Filho, H. C., Martinatto, C. da S., Schulz, L. R., Silva, J. P. de S. & Consoli, N. C. (2023). Mechanical and microstructural response of iron ore tailings under low and high pressures considering a wide range of molding characteristics. *Mining, 3*, 712–730. https://doi.org/10.3390/mining3040039.

Buikema, N. D. (2015). *Stabilization of iron mine tailings through microbially induced calcite precipitation*. Michigan Technological University.

Carmignano, O. R., Vieira, S. S., Teixeira, A. P. C., Lameiras, F. S., Brandão, P. R. G. & Lago, R. M. (2021). Iron ore tailings: Characterization and applications. *Journal of the Brazilian Chemical Society, 32(10)*, 1895-1911. https://dx.doi.org/10.21577/0103-5053.20210100.

Carmo, F. F. do, Kamino, L. H. Y., Junior, R. T., Campos, I. C. de, Carmo, F. F. do, Silvino, G., Castro, K. J. da S. X. de, Mauro, M. L., Rodrigues, N. U. A., Miranda, M. P. de S. & Pinto, C. E. F. (2017). Fundão tailings dam failures: The environment tragedy of the largest technological disaster of Brazilian mining in global context. *Perspectives in Ecology and Conservation*, *15*(3), 145–151. https://doi.org/10.1016/j.pecon.2017.06.002.

Carneiro, A. D. A. (2020). *Comportamento mecânico de um rejeito de minério de ferro estabilizado com polímero e do compósito rejeito-polímero reforçado com fibras de polipropileno*. Universidade de Brasília.

Carvalho, J. V. de A., Wagner, A. C., Scheuermann Filho, H. C., Chaves, H. M., Silva, J. P. S., Delgado, B. G. & Consoli, N. C. (2023). Evaluation of strength parameters for application in cemented iron ore tailings stacks. *Indian Geotechnical Journal, 53*, 775–788. https://doi.org/10.1007/s40098-023-00712-9.

Chaves, H. M. (2023). *Estudo geotécnico de uma pilha experimental de misturas compactadas de rejeito de minério de ferro filtrado - cimento Portland*. Universidade Federal do Rio Grande do Sul.

Consoli, N. C., Bittar Marin, E. J., Quiñónez Samaniego, R. A., Heineck, K. S. & Johann, A. D. R. (2019). Use of sustainable binders in soil stabilization. *Journal of Materials in Civil Engineering*, *31*(2), 1–7. https://doi.org/10.1061/(asce)mt.1943-5533.0002571.

Consoli, N. C., Leon, H. B., da Silva Carretta, M., Daronco, J. V. L. & Lourenço, D. E. (2019). The effects of curing time and temperature on stiffness, strength and durability of sand-environment friendly binder blends. *Soils and Foundations*, *59*(5), 1428–1439. https://doi.org/10.1016/j.sandf.2019.06.007.

Consoli, N. C., Vogt, J. C., Silva, J. P. S., Chaves, H. M., Filho, H. C. S., Moreira, E. B. & Lotero, A. (2022). Behaviour of compacted filtered iron ore tailings–portland cement blends: new brazilian trend for tailings disposal by stacking. *Applied Sciences*, *12*(2), 836. https://doi.org/10.3390/app12020836.

Dauce, P. D., Castro, G. B. De, Lima, M. M. F. & Lima, R. M. F. (2019). Characterisation and magnetic concentration of an iron ore tailings. *Journal of Materials Research and Technology*, *8*(1), 1052–1059. https://doi.org/10.1016/j.jmrt.2018.07.015.

Davila, R. B., Fontes, M. P. F., Pacheco, A. A. & Ferreira, M. da S. (2020). Heavy metals in iron ore tailings and floodplain soils affected by the Samarco dam collapse in Brazil. *Science of the Total Environment*, *709*, 136151. https://doi.org/10.1016/j.scitotenv.2019.136151.

De Jesus, C. A. G. & Joaquim, L. G. (2020). Sumário Brasileiro Mineral 2020: Ferro. *Agência Nacional de Mineração*, 1–3.

Edraki, M., Baumgartl, T., Manlapig, E., Bradshaw, D., Franks, D. M. & Moran, C. J. (2014). Designing mine tailings for better environmental, social and economic outcomes: A review of alternative approaches. *Journal of Cleaner Production*, *84*(1), 411–420. https://doi.org/10.1016/j.jclepro.2014.04.079.

Farenzena, H. P., Bruschi, G. J., Medina, G. S., Silva, J. P. S., Lotero, A. & Consoli, N. C. (2024). Iron ore tailings stabilization with alternative alkali-activated cement for dry stacking: mechanical and microstructural insights. *Canadian Goetechnical Journal*, 1–19. https://doi.org/10.1139/cgj-2023-0125

Ferrazzo, S. T., Araújo, M. T. de, Bruschi, G. J., Chaves, H. M., Korf, E. P. & Consoli, N. C. (2023). Mechanical and environmental behavior of waste foundry sand stabilized with alkali-activated sugar cane bagasse ash-eggshell lime binder. *Construction and Building Materials*, *383*, 131313. https://doi.org/10.1016/j.conbuildmat.2023.131313.

Fontes, W. C., Franco de Carvalho, J. M., Andrade, L. C. R., Segadães, A. M. & Peixoto, R. A. F. (2019). Assessment of the use potential of iron ore tailings in the manufacture of ceramic tiles: From tailings-dams to "brown porcelain." *Construction and Building Materials*, *206*, 111–121. https://doi.org/10.1016/j.conbuildmat.2019.02.052.

Fracaro, T. P. (2022). *Análise do comportamento mecânico de rejeito de minério de ferro submetido a ativação alcalina pelo método "two parts"*. Universidade Federal do Rio Grande do Sul.

Freitas, C. M. & Da Silva, M. A. (2019). Work accidents which become disasters: Mine tailing dam failures in Brazil. *Revista Brasileira de Medicina Do Trabalho*, *17*(1), 21–29. https://doi.org/10.5327/Z1679443520190405.

Garcia-Lodeiro, I., Palomo, A. & Fernández-Jiménez, A. (2015). An overview of the chemistry of alkali-activated cement-based binders. In *Handbook of Alkali-Activated Cements, Mortars and Concretes* (pp. 19–47). Woodhead Publishing. https://doi.org/10.1533/9781782422884.1.19.

Gentil, G. A. (2020). *Avaliação da estabilização química de um rejeito de minério de ferro para aplicação em infraestrutura de pavimentos*. Universidade Federal de Ouro Preto.

Huntzinger, D. N. & Eatmon, T. D. (2009). A life-cycle assessment of Portland cement manufacturing: comparing the traditional process with alternative technologies. *Journal of Cleaner Production*, *17*(7), 668–675. https://doi.org/10.1016/j.jclepro.2008.04.007.

Jaskulski, F. M. (2022). *Análise do comportamento mecânico de rejeito de minério de ferro submetido a ativação alcalina pelo método "one-part"*. Universidade Federal do Rio Grande do Sul.

Kossoff, D., Dubbin, W. E., Alfredsson, M., Edwards, S. J., Macklin, M. G. & Hudson-Edwards, K. A. (2014). Mine tailings dams: Characteristics, failure, environmental impacts, and remediation. *Applied Geochemistry*, *51*, 229–245. https://doi.org/10.1016/j.apgeochem.2014.09.010.

Levandoski, W., Ferrazzo, S. T., Bruschi, G. J., Consoli, N. C. & Korf, E. P. (2023). Mechanical and microstructural properties of iron mining tailings stabilized with alkali-activated binder produced from agro-industrial wastes. *Scientific Reports, 13*, 15754. https://doi.org/10.1038/s41598-023-42999-x.

Machado, L. da S. (2022). *Estabilização de rejeitos de minério de ferro filtrados com uso de aditivos para disposição em pilhas compactadas (dry stacking)*. Universidade Federal do Rio Grande do Sul.

Mafessoli, M. (2022). *Comportamento mecânico de um rejeito de minério deferro estabilizado com cimento Portland*. Universidade Federal do Rio Grande do Sul.

Mafessoli, M., Marques, S. F. V., Scheuermann Filho, H. C. & Consoli, N. C. (2023). Response of artificially cemented iron ore tailings for dry stacking disposal over a wide range of stresses. *Indian Geotechnical Journal, 53*, 904–915. https://doi.org/10.1007/s40098-023-00711-w.

Manjunatha, L. S. & Sunil, B. (2013). Stabilization/solidification of iron ore mine tailings using cement, lime and fly ash. *International Journal of Research in Engineering and Technology*, *02*(12), 625–635. https://doi.org/10.15623/ijret.2013.0212107.

Maruthupandian, S., Chaliasou, N. A. & Kanellopoulos, A. (2021). Recycling mine tailings for a sustainable future built environment. In *Energy and Sustainable Futures* (pp. 163–169). Springer. https://doi.org/10.1007/978-3-030-63916-7_21.

MME. (2019). *Boletim do Setor Produção Mineral Comércio Exterior. 28.*

Moretti, J. P., Sales, A., Almeida, F. C. R., Rezende, M. A. M. & Gromboni, P. P. (2016). Joint use of construction waste (CW) and sugarcane bagasse ash sand (SBAS) in concrete. *Construction and Building Materials*, *113*, 317–323. https://doi.org/10.1016/j.conbuildmat.2016.03.062.

Pan, Z., Zhang, C., Li, Y. & Yang, C. (2022). Solidification/stabilization of gold ore tailings powder using sustainable waste-based composite geopolymer. *Engineering Geology*, *309*, 106793. https://doi.org/10.1016/j.enggeo.2022.106793.

Pelisser, G., Ferrazzo, S. T., Mota, J. D., dos Santos, C. P., Pelisser, C., Rosa, F. D. & Korf, E. P. (2023). Rice husk ash-carbide lime as an alternative binder for waste foundry sand stabilization. *Environmental Science and Pollution Research*, *30*(14), 42176–42191. https://doi.org/10.1007/s11356-022-25117-8.

Provis, J. L. & Van Deventer, J. S. J. (2014). Binder chemistry – Blended systems and intermediate Ca content. In *Alkali Activated Materials: RILEM State-of-the-Art Reports* (pp. 125–144). Springer. https://doi.org/10.1007/978-94-007-7672-2_5.

Reis, J. B., Pelisser, G., Levandoski, W. M. K., Ferrazzo, S. T., Mota, J. D., Silveira, A. A. & Korf, E. P. (2022). Experimental investigation of binder based on rice husk ash and eggshell lime on soil stabilization under acidic attack. *Scientific Reports*, *12*(1), 1–11. https://doi.org/10.1038/s41598-022-11529-6.

Saldanha, R. B., Caicedo, A. M. L., de Araújo, M. T., Scheuermann Filho, H. C., Moncaleano, C. J., Silva, J. P. S. & Consoli, N. C. (2023). Potential use of iron ore tailings for binder production: A life cycle assessment. *Construction and Building Materials*, *365*, 130008. https://doi.org/10.1016/j.conbuildmat.2022.130008.

Servi, S., Lotero, A., Silva, J. P. S., Bastos, C. & Consoli, N. C. (2022). Mechanical response of filtered and compacted iron ore tailings with different cementing agents: Focus on tailings-binder mixtures disposal by stacking. *Construction and Building Materials*, *349*, 128770. https://doi.org/10.1016/j.conbuildmat.2022.128770.

Tuck, B. C. A. & Ghalayini, Z. T. (2016). Iron Ore. In *U.S. Geological Survey Minerals Yearbook* (pp. 1–13).

Wang, Y., Wen, Z., Cao, X., Zheng, Z. & Xu, J. (2020). Environmental efficiency evaluation of China's iron and steel industry: A process-level data envelopment analysis. *Science of the Total Environment*, *707*, 135903. https://doi.org/10.1016/j.scitotenv.2019.135903.

Ye, N., Yang, J., Liang, S., Hu, Y., Hu, J., Xiao, B. & Huang, Q. (2016). Synthesis and strength optimization of one-part geopolymer based on red mud. *Construction and Building Materials*, *111*, 317–325. https://doi.org/10.1016/j.conbuildmat.2016.02.099.

Yellishetty, M., Ranjith, P. G. & Tharumarajah, A. (2010). Iron ore and steel production trends and material flows in the world: Is this really sustainable? *Resources, Conservation & Recycling, 54*(12), 1084–1094. https://doi.org/10.1016/j.resconrec.2010.03.003.

Zahmak, A., Abdallah, M., Jarah, B. & Arab, M. G. (2021). Environmental performance of alkali-activated binders for ground improvement. *Transportation Geotechnics*, *31*, 100631. https://doi.org/10.1016/j.trgeo.2021.100631.

Chapter 11

Leaching Behavior of Alkali-Activated Materials in Environmental Geotechnics

Suéllen Tonatto Ferrazzo[1,*]
Giovani Jordi Bruschi[1]
and Eduardo Pavan Korf[2]
[1]Graduate Program in Civil Engineering, Universidade Federal do Rio Grande do Sul, Porto Alegre, RS, Brazil
[2]Graduate Program in Environmental Science and Technology, Universidade Federal da Fronteira Sul, Erechim, RS, Brazil

Abstract

Assessing the environmental risks of alkali-activated materials (AAM) produced from mining tailings, agroindustrial and industrial wastes through metal leaching is essential for determining their technical feasibility. This chapter presents an overview of the leaching behavior of alkali-activated materials in environmental geotechnics, discussing results and testing methods. The leaching behavior of AAM focuses on agroindustrial and industrial solid waste, natural soils, metals contaminated-soils and mining tailings. In general, the alkaline activation technique enabled the efficient stabilization/solidification of AAM, resulting in materials without toxicity characteristics, with some meeting requirements for water quality and earth construction. Optimization of alkaline cement in relation to environmental performance, strategies/dosages of alkaline cement applied to the stabilization of waste, soil or mining tailings, and understanding the inertization mechanisms of metals considered problematic can still be further

* Corresponding Author's Email: suellenferrazzo@hotmail.com.

In: Environmental Geotechnics
Editors: E. Pavan Korf, P. Domingos Marques Prietto, C. Müller et al.
ISBN: 979-8-89113-651-9

explored. The leaching methods used in these research fall into the categories of static tests and dynamic tests, subdivided into batch tests, column tests and tests on monolithic samples. The size of the particles of the materials, the leaching solution and the duration of the tests varied according to the chosen method. Except for the column and monolithic tests, specimens were segregated to sizes of 3 cm, 9.5 mm, 4 mm, 3 mm or 1 mm. As an extracting solution, distilled water or an acidic solution was applied. The duration of the tests ranged from 10 hours to two months. The different tests provide specific estimates of leaching behavior of AAM, making it difficult to compare results from the literature and also pointing to the need for international standardization of leaching tests for AAM. It is expected that this information will serve as a reference for new studies in AAM, aiming to overcome the primary barrier for applications, i.e., ensuring that the material poses no environmental risks.

Keywords: geotechnical materials, alkali-activated cement, metal encapsulation, leaching tests

Alkali Activated Materials in Environmental Geotechnics

In the field of environmental geotechnics, research into the stabilization of materials using alkaline cements, including alkali-activated binders and geopolymers, is primarily focused on investigating their mechanical behavior and microstructure.

In recent years, there has been a growing interest in assessing the leaching of contaminants, particularly metals. This is crucial due to the use of waste materials as precursors, which may contain contaminants in their chemical composition. Therefore, evaluating metal leaching helps comprehend the environmental risks associated with materials cemented by alkaline activation over the long term and provides technical support for potential applications in engineering projects.

However, there is no internationally standardized regulation for testing leaching in alkali-activated materials (AAM). Consequently, various methods from numerous regulations have been employed, most of which were originally designed for characterizing and classifying solid waste. In this context, it becomes imperative to understand the outcomes yielded by different approaches for evaluating leaching in AAM within geotechnical contexts. Figure 1 provides an overview of research in this area, covering the types of materials used in stabilization and leaching tests.

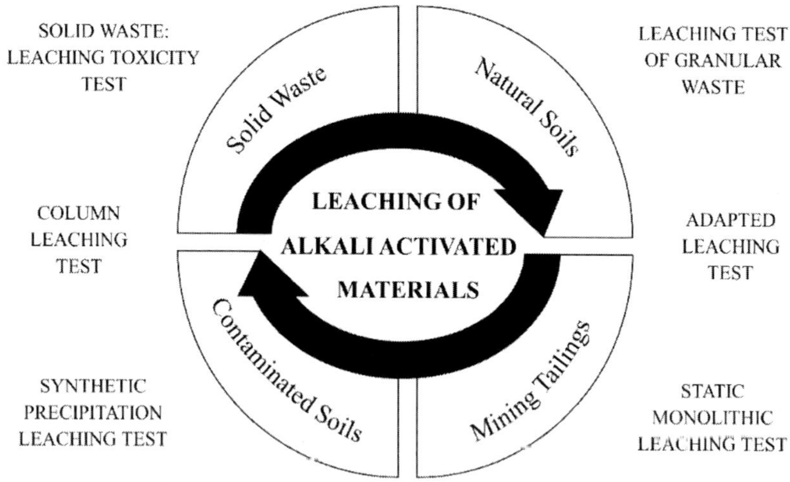

Figure 1. Overview of leaching from alkali-activated materials in environmental geotechnics.

The assessment of metal encapsulation in AAM primarily encompasses industrial and agro-industrial solid wastes, natural soils, contaminated soils, and mining tailings. The leaching methods employed are quite diverse and can be categorized into batch tests, percolation tests, and tests involving monolithic samples. Subsequently, the literature is discussed, highlighting the contributions of each study towards advancing the knowledge and leaching tests.

Metal Leaching from Alkali-Activated Materials

Solid Waste

In the literature, there are several studies on the leaching of metals from various AAM, either as a complement to mechanical performance evaluation or mainly focused on environmental performance. In this context, Hoy et al. (2016) assessed the leaching of metals from geopolymers produced from reclaimed asphalt pavement (RAP) and fly ash using the Toxicity Characteristic Leaching Procedure (TCLP) test - method 1311 (USEPA, 1992). Hoy et al. (2016) indicated that the concentrations of metals such as As, Ba, Cd, Cr, Pb, Hg, Se, and Ag did not exceed the maximum limits established by drinking water quality regulations (USEPA, 1999) and hazardous waste regulations (EPA-Victoria, 2009). This indicates that geopolymers pose no environmental risks for application in pavement bases.

Regarding their use in flexible pavement layers, Pelissaro et al. (2023) developed an alkali-activated metakaolin-based cement (precursor) with rice husk ash and sodium hydroxide (an alternative alkaline activator) to stabilize RAP (along with stone powder and gravel). Through the NBR 10005 test (ABNT, 2004), the assessment of leaching behavior revealed that mixtures with 40% and 50% RAP showed no toxicity and met the water quality limits set by USEPA (USEPA, 2022). The leaching results from Pelissaro et al. (2023) demonstrate that RAP-cemented materials pose no environmental risks, which, combined with satisfactory mechanical performance, provide technical support for application in pavement bases. The stabilization of RAP with alkali-activated cement is highly promising, considering the need for waste reuse in road construction projects in Brazil.

As solid waste generated in metal industries, slag has been incorporated into waste management approaches to reduce environmental risks and add value (Movaheditabar et al. 2022). One such approach is the incorporation of slag as a raw material in the production of alkali-activated cements. Wang et al. (2018) identified that after alkaline activation, nickel-iron slag exhibited a significant reduction in the leaching of Mg, Cr(VI), and Cr(III) (TCLP test), transforming the hazardous waste into a non-toxic construction material. Mg immobilization occurred through encapsulation, sorption, surface complexation, substitution, and precipitation. Authors also noted that the leaching of Cr(VI) was not governed by chemical precipitation, whereas Cr(III) precipitated as $Cr(OH)_3$. The stabilization of Cr in different oxidation states within the AAM matrix is an excellent result in terms of environmental performance, given the toxicity and DNA-damaging properties of chromium, with Cr(VI) being more toxic than Cr(III) and considered a potent carcinogen (Desmarais & Costa, 2019). Thus, the results from Wang et al. (2018) overcome the primary barrier (environmental risk) to the valorization of nickel-iron slag, which is a significant concern due to heavy metals, leaching potential, environmental pollution, low recycling rates, and high landfill disposal (Ha et al. 2023; Han & Hong, 2018; Wang et al. 2018).

Another type of slag used as a precursor for cements is fayalite slag. This is an iron-rich waste from nickel and copper manufacturing, typically disposed of in landfills or used in low-value applications (Adediran et al. 2021). Adediran et al. (2021) evaluated the leaching of alkali-activated binders made from granulated fayalite slag with 28 days of curing, following the waste classification standard EN12457 (EN, 2002). Comparatively, to the concentrations of metals in the leachate extracts from the precursor materials, there was an increase in the leaching of Cr, Mo, Ni, and Se from the alkali-activated pastes due to the high alkalinity conditions during the tests. Another important result is that the alkali-activated binder leached metals at concentrations lower than the maximum limits established for use in paved structures in Finland [Decree 843 (FINLEX, 2017)], indicating the technical feasibility in terms of environmental risks.

The hydrometallurgical processing of zinc from sulfide minerals generates residues, such as iron-rich slag, which require careful disposal due to their potential to contaminate soil and groundwater (Ponomar et al. 2023; Tang et al. 2018). With the goal of establishing a safe strategy for using this waste, Ponomar et al. (2023) investigated the leaching behavior of alkali-activated zinc slag. Leaching tests were conducted on the AAM following the procedures outlined in the EN12457 standard (EN, 2002). It was observed that alkaline activation of the slag results in increased leaching of certain metals (e.g., As, Mo, Cr, Sb, Se, V). However, it also enables the immobilization of Ba, Pb, and Zn due to the incorporation of these elements into the cementitious gels. Additionally, it was noted that liquid silica in the alkaline activator solutions and a higher water content are factors contributing to reduced leaching of the AAM. Compared to the limits for utilizing waste in earthworks in Finland [Decree 843 (FINLEX, 2017)], the concentrations of Pb, Cu, and Zn leached exceeded these limits, while As, Cr, and Ni met the requirements, and Mo, Sb, and V still pose challenges for alkali-activated slag applications. Ponomar et al. (2023) recommend future research to investigate the mechanisms of immobilization of specific metals both before and after alkaline activation of the waste. Thus, advancing the immobilization of problematic metals is essential for the technical viability of AAM applications and represents a research gap to be explored.

Building upon the method proposed by Garcia-lodeiro et al. (2021), Shagñay et al. (2023) evaluated the leaching of alkali-activated pastes based on blast furnace slag by immersing 3 cm samples in a 6M NH_4NO_3 solution (i.e., a leaching kinetics catalyst) for 28 days. Authors indicated that the AAMs developed exhibited greater resistance to decalcification compared to Portland cement due to the absence of portlandite and a high degree of silicate chain polymerization. It was also observed that the higher resistance of the AAM is associated with its less porous microstructure. The test conducted by Shagñay et al. (2023) corresponds to an accelerated leaching process to assess primarily the leaching of Ca, which is a distinct purpose from the aforementioned research. While some studies focus on the stabilization/solidification of contaminants, that of Shagñay et al. (2023) is oriented towards microstructural changes and the consequent durability of alkali-activated cement, demonstrating the numerous insights that can be gained from leaching tests in AAM.

Fly ash from municipal solid waste incineration is a waste that requires proper treatment to prevent secondary pollution due to the presence of heavy metals and other contaminants in its composition (Xu et al. 2019). A way to stabilize/solidify this fly ash is through alkaline activation. Xu et al. (2019) explored the leaching of metals from alkali-activated fly ash from municipal solid waste incineration through various batch tests (i.e., waste characterization tests). Tests were conducted following the procedures outlined in the following standards: HJ/T 299 (MEE, 2007b) and HJ/T 300 (MEE, 2007c) (Chinese standards for evaluating waste toxicity); method 1311 – TCLP test (USEPA, 1992); method 1312 – SPLP test (USEPA, 1994) (simulating acid rain environments); and CEN/TS 14429 (CEN, 2015) (waste characterization - assessing the pH influence on leaching). The results of the first four tests showed that the leachates had different pH values, which directly influenced metal leaching. For example, Cu, Cr, and Pb leached in acidic environments with a pH<6. Authors suggest that the CEN/TS 14429 standard (CEN, 2015) is more comprehensive in evaluating the leaching behavior of metals under different pH conditions, as it covers both acidic and alkaline environments (pH values in the range of 4-12). Xu et al. (2019) study highlights the variability in the leaching behavior of AAM depending on the specifics of each of the methods applied and emphasizes the importance of considering pH as a significant factor.

The primary by-product of the metal foundry industry, waste foundry sand (WFS), can be fully reused through cementation and compaction for various applications in geotechnical engineering, such as retaining walls, pipeline beds, road bases, and sub-bases, etc. (Ferrazzo et al. 2023) (see Chapter 6). Ferrazzo et al. (2023a) investigated the stabilization of WFS using an alkali-activated binder (AAB) derived from sugarcane bagasse ash and eggshell lime, in comparison to the use of Portland cement (Type III). The authors conducted leaching tests on the WFS-binder mixtures following the NBR 10005 method (ABNT, 2004) and compared the results with toxicity and water quality standards from Brazilian and international regulations (CONAMA, 2013; USEPA, 2022; VROM, 2000). The WFS-AAB matrices after 28 days of curing demonstrated the ability to encapsulate Cd, Cr, Mn, Pb, and Zn present in the waste due to the formation of sodium aluminosilicate hydrate (N-A-S-H) gel. The leachate extracts from the WFS-AAB and WFS-Portland cement mixtures showed no toxicity. Thus, it was found that AAB can replace traditional cement in the treatment of WFS.

Ferrazzo et al. (2023b) evaluated the leaching behavior of WFS stabilized with AAB from agroindustrial wastes through batch [NBR 10005 (ABNT, 2004)] and column tests [D4874 (ASTM, 2014)]. WFS-Portland cement (PC) was evaluated as a control group. All WFS-AAB and WFS-PC mixtures after 7 days of curing presented no metal toxicity. The WFS-AAB matrices were effective in encapsulating Cd, Cr and Pb present in the WFS and SCBA compositions. In addition, batch test results were more favorable than column test concerning water quality limits (CONAMA, 2013; USEPA, 2022; VROM, 2000). According to Ferrazzo et al. (2023b), binder type, metals leaching patterns, and leaching methods are key factors for assessing the leaching behavior of WFS stabilized with traditional and alkaline cements. This study demonstrated that there are several factors involved in evaluating the leaching behavior of AAM, highlighting the importance of using different leaching tests.

Metals-Contamined Soils

Metal contamination on soils is a global issue, and among the treatment techniques, stabilization/solidification stands out as one of the most sought-after, bringing improvements to strength and leaching properties (Jiang et al. 2022; Liu et al. 2017). This treatment can be cost-effective and performance-enhanced when utilizing industrial waste as stabilizing agents (Jiang et al. 2022). In this context, the use of these types of waste in the production of alkali-activated cements for the treatment of soils contaminated with different metals is noteworthy.

The stabilization/solidification of Pb-contaminated soil using a geopolymer based on slag and fly ash was evaluated by Li et al. (2019). Leaching tests using Synthetic Precipitation Leaching Procedure (SPLP), following method 1312 (USEPA, 1994), revealed that the concentration of leached Pb decreased with an increase in the geopolymer content, which proved to be more efficient than Ordinary Portland Cement. The addition of 10% geopolymer reduced the concentration of leached Pb in the soil to meet the maximum limits for hazardous waste in China [GB5085 (MEE, 2007a)]. Meanwhile, the incorporation of 20% geopolymer complied with the standards for surface water quality for agricultural use in the country [GB3838 (MEE, 2002)]. Given the significant problem of Pb-contaminated soils in China (Luo et al. 2023; Wei et al. 2015) and the toxicity of this heavy metal (Kumar et al. 2020; Wani et al. 2015), the results of the aforementioned study can assist in decision-making regarding industrial waste management processes and the remediation of contaminated areas in the country.

The treatment of soil contaminated with Pb, Zn, and Cd using a potassium monopotassium phosphate-activated slag binder was investigated by Feng et al. (2020). Leaching tests [Toxicity Characteristic Leaching Procedure - TCLP (USEPA, 1992)] revealed that heavy metals were immobilized through various mechanisms: precipitation of hydroxides, ion exchange with hydroxyapatite, and adsorption and encapsulation by calcium silicate hydrate. It was observed that alkaline activation reduced the leaching of metals in the soil mixture with 15% slag, particularly with a gradual increase in the activator/precursor ratio. Authors found that activator/precursor ratios of 1:50, 1:25, and 1:12.5 resulted in the minimum remediation target for Pb, Zn, and Cd. Feng et al. (2020) identified metal immobilization mechanisms based on pH measurements, specific surface area analyses, and X-ray diffraction (XRD).

Pu et al. (2022) assessed the environmental behavior of Pb-contaminated soil stabilized with a silico-aluminophosphate geopolymer through leaching tests, following HJ/T 299 (MEE, 2007b). Soil-geopolymer specimens containing powdered alkaline activator exhibited higher Pb^{2+} leaching at curing times of 7 and 14 days. However, at 28 days, the state (solid or liquid) of the alkaline activator had no influence on the leaching behavior. The application of 8% geopolymer in soil with up to 2% Pb^{2+} resulted in metal leaching concentrations below the toxicity limit for hazardous waste in China (MEE, 2007a). Furthermore, it was observed that the addition of 8% Ordinary Portland Cement achieved greater reduction in Pb^{2+} leaching but exhibited lower resistance to acidic and alkaline environments compared to the use of the geopolymer. According to Pu et al. (2022), the developed alkali geopolymer has the potential for application in the treatment of contaminated soils under extreme acidic conditions. Notably, the study included pH and electrical conductivity determinations and responses from scanning electron microscopy (SEM) and XRD techniques, corroborating the leaching results, which is important for reinforcing the conclusions.

Recently, Komaei et al. (2023) studied the treatment of silty sand contaminated with As through stabilization/solidification using alkali-activated ground granulated blast furnace slag. The results of the TCLP test revealed a reduction of over 90% in As leaching from the soil with 20% binder after 28 days of curing (23°C). This behavior was associated with the formation of albite and anorthite crystals, as well as the identification of sodium-calcium aluminosilicate hydrate [(N, C)–A–S–H] and calcium silicate hydrate (C–S–H) gels through XRD and SEM analyses. Additionally, the authors observed similar As leaching behavior between soil treated with AAB and Ordinary Portland Cement. Due to the carcinogenic potential and bioaccumulation of As, soils contaminated with this heavy metal pose risks to the environment and public health (Rahman et al. 2023). In the face of this global issue, Komaei et al. (2023) made an important contribution by proposing an effective remediation strategy. Through the alkaline activation of an industrial byproduct, it was possible to immobilize As in the contaminated soil and also recycle the slag.

Natural Soils and Mining Tailings

In addition to mechanical behavior, the leaching of metals from natural soils and mining waste stabilized with alkali cements has become a subject of investigation. Tigue et al. (2018) conducted column leaching tests to investigate the encapsulation of metals in a sandy-silty soil stabilized with coal fly ash-based geopolymer after 28 days of curing. Both the soil and fly ash were sourced from a local power plant. Significant leaching of Al, Na, and Si from the cemented matrix was observed, as these metals are abundant in the chemical composition of the raw materials and alkaline activator (Na_2SiO_3+NaOH solution). The lower dissolution of Fe was attributed to the fixation of this metal in the geopolymer gel. Leached Ca, originating from the fly ash and soil, occurred due to the insufficient amount of reactive silica and alumina to promote cementitious gel formation. Except for Cd and Pb, there was a reduction in metal dissolution over time. Additionally, as exhibited higher mobility in alkaline environments compared to other heavy metals. However, the concentrations of leached metals in this study were not compared to maximum toxicity or water quality limits, making it difficult to understand the environmental risks of the soil-geopolymer mixture.

Cristelo et al. (2021) investigated the leaching of contaminants from clay stabilized with an alkali-activated binder (AAB) made from fly ash and glass waste (precursors) and alkaline solution recycled from aluminum extrusion (alternative alkaline activator) after 28 days of curing. The leaching test was conducted according to the method established by EN12457 (EN, 2002). The concentrations of contaminants in the leachates were compared to the limits set by Council Decision 2003/33/CE for disposal in inert and non-hazardous waste landfills (COUNCIL-EU, 2003). The results showed that the concentrations of leached Pb, Zn, and sulfates pose no contamination risks, as they were below the limits for hazardous waste. Pb, present in the glass waste, leached at slightly higher concentrations but still remained below the maximum limit for inert waste.

Regarding mining waste treatment techniques, Pereira dos Santos (2021) evaluated the leaching behavior of gold ore tailings stabilized with an AAB derived from sugarcane bagasse ash and carbide lime, using a column test according to ASTM D4874 (ASTM, 2014). The author found that the waste-binder matrix effectively encapsulated metals (e.g., Al, Mn, Ba, Cd, Cu, Cr, and Zn) in compliance with water quality standards from various regulations (CONAMA, 2013; USEPA, 2022; VROM, 2000). Furthermore, the AAB proved to be more efficient in immobilizing metals in the gold ore tailings mixture compared to Ordinary Portland Cement (Type III).

Zhang et al. (2022) investigated the leaching of a geopolymer developed from alkali-activated gold ore tailings. Static monolithic leaching tests following method 1315 (USEPA, 2017) were conducted to assess the release of metals from the alkali-activated material when immersed in natural water, saline water, and moderately acidic water (pH 4.5). It was observed that the transfer of metals to the liquid phase was not closely associated with the immersion solutions. Overall, the geopolymer efficiently immobilized metals such as Cu, Mn, Mo, and Se present in the tailings. However, high concentrations of As, Si, and S were leached due to geopolymerization and the dissolution of arsenic-based compounds. According to the authors, the leaching behavior of the alkali-activated material based on mining waste was governed by diffusion and surface wash-off mechanisms. Additionally, Zhang et al. (2022) proposed a first-order reaction/diffusion model capable of reasonably predicting the leaching kinetics of metals from the alkali-activated material.

The stabilization of gold ore tailings with a "sustainable" composite geopolymer was evaluated by Pan et al. (2022). The alkali cement was produced by mixing precursors, including fly ash, gray-black powder, blast furnace slag, and metakaolin. Commercial sodium hydroxide was used as the alkaline activator, while calcium carbide residue and gypsum powder served as calcium sources. In this study, metal immobilization was investigated through leaching tests following the horizontal vibration method specified in HJ/T 299 (MEE, 2007b). After 7 days of curing, it was found that all evaluated metals (As, Pb, Zn, Ni, Cr, and Co) were immobilized in the cemented matrices, meeting surface water quality standards [GB3838 (MEE, 2002)]. As, a metal with complex immobilization, present in the tailings, was immobilized through precipitation, physical encapsulation, adsorption, and ion exchange (Pan et al. 2022).

The results of the studies by Pereira dos Santos (2021), Zhang et al. (2022), and Pan et al. (2022) are highly attractive from an environmental perspective for the proper management of gold ore tailings, as they are complex due to the presence of sulfides, which can oxidize and form sulfuric acid, and numerous heavy metals susceptible to leaching under acidic conditions (Ahmari & Zhang, 2013; Kiventerä et al. 2018). Moreover, the alkaline cements developed by Pereira dos Santos (2021) and Pan et al. (2022) incorporated various industrial and agro-industrial residues as precursors and calcium sources, demonstrating that these residues, usually discarded in landfills, have potential applications and added value.

Furthermore, bauxite mining residues occupy large areas of land and have negative impacts on terrestrial resources and the environment, so finding ways to valorize this waste is essential (Ou et al. 2022). Bruschi et al. (2022) investigated the stabilization of bauxite residue with an AAB derived from waste materials (sugarcane bagasse ash and carbide lime) using a leaching test adapted from ASTM D559 (ASTM, 2015) and NEM7375 (NEN, 2004) standards. The first standard corresponds to the wetting and drying cycle durability method, and the latter regulates the leaching test of monolithic materials. The adaptation involved immersing the specimens in distilled water during the durability test, and after every three wetting and drying cycles, the liquid was collected. The authors found that the bauxite residue-AAB mixture showed good capability to encapsulate Al, Cr, and Se, meeting water quality limits, while Type III Portland cement was more effective in stabilizing Cd and Hg. According to the authors, the leaching behavior of the waste-binder matrices was controlled by binder content, porosity, and pH of the environment. The significant influence of the porosity of the cemented matrices on the concentrations of leached metals was associated with the leaching test being conducted on cylindrical specimens parallel to wetting and drying cycles.

In the context of the final disposal of mining waste, alkali-activated cements applied to mining residues result in satisfactory mechanical behavior and lower environmental impacts compared to Ordinary Portland Cement and have potential for use in the dry stacking method (Saldanha et al. 2023; Servi et al. 2022). However, it is also necessary to investigate the leaching behavior of contaminants (see Chapter 11). In this regard, the research results of Zhang et al. (2022), Pereira dos Santos (2021), Pan et al. (2022), and Bruschi et al.

(2022) are important for enhancing the competitiveness of alkaline cements based on waste compared to traditional cement and contribute to developing solutions for the sustainable management of mining residues.

Methods and Procedures for Assessing Leaching

Leaching tests allow estimating the leaching behavior of organic or inorganic elements from waste and solid materials in different management scenarios (final disposal and uses). In the aforementioned studies, various leaching tests were used to evaluate the environmental performance of AAM, which is summarized in Table 1. It is worth noting that the analysis of metal leaching has been carried out using numerous methods categorized into static tests (non-renewed leaching fluid) and dynamic tests (renewed leaching fluid), further divided into batch tests, column tests (percolation system), and tests for monolithic samples.

Batch tests, represented by Method 1311 (TCLP), Method 1312 (SPLP), EN12457, NBR 10005, HJ/T 299, HJ/T 300, and CEN/TS 14429, are the most used to assess the leaching behavior of AAM. These tests were developed for the characterization/classification of solid waste and have some variations in their procedures, except for TCLP and NBR 10005, which are very similar. In general, batch tests belong to the category of leaching tests in which the specimen is placed in a certain volume of leaching fluid with agitation (in most methods) for a specified period (Siddique et al. 2010). Batch tests aimed at evaluating the toxicity characteristics of waste simulate the leaching of contaminants under landfill disposal conditions, using some organic acid (Siddique et al. 2010; Xu et al. 2019). In landfills, organic acids are formed from the decomposition of organic matter in the waste. The SPLP method (USEPA, 1994) differs from the others as it is applied to determine the mobility of organic and inorganic analytes in waste, as well as in liquids and soils. This test simulates the exposure of waste or material to acid rain to assess leaching and potential environmental damage (Feng et al. 2020; Xu et al. 2019).

According to Feng et al. (2020), TCLP and SPLP are recommended methods by the U.S. environmental agency for assessing the toxicity characteristics of solid materials, as well as solid waste, including soils, sediments, and construction materials. Batch tests, such as those exemplified above, have a simple design, low cost, and are conducted over a short period (Feng et al. 2020; Naka et al. 2016). However, their procedures encompass a limited range of conditions, e.g., pH, redox potential, extraction time, solid/liquid ratio, etc. (Feng et al. 2020). Xu et al. (2019) believe that the pH-dependent method - CEN/TS 14429 (CEN, 2015) is more comprehensive for assessing the leaching behavior of metals in environments with different pH conditions, as the initial addition of acid/base conditions a wide pH range (4–12). The authors recommend the CEN/TS 14429 (CEN, 2015) method for evaluating the leaching behavior of recycled products and materials incorporating hazardous industrial waste.

Percolation systems, represented by column leaching tests, were initially developed to simulate the percolation of groundwater through a granular bed, with upward or downward and continuous or intermittent flow, under saturated or unsaturated conditions (Siddique et al. 2010), and under different hydraulic gradients depending on the geotechnical characteristics of the material [D5084 (ASTM, 2016)]. This category of leaching tests provides a closer representation of real field conditions and an understanding of the long-term transfer of contaminants from soil to groundwater. However, these tests can be slow, costly, and more complex compared to batch tests (Naka et al. 2016). AAM may have very low porosities and microstructures that require long periods of time to assess leaching through percolation systems, respecting the recommended conditions and hydraulic gradients. Column tests, as described in D4874 (ASTM, 2014), aim to maximize the leaching of metals from a solid material by passing a fluid, usually distilled/deionized water, upwards through a saturated medium (Siddique et al. 2010). Upflow tests aim for an equal distribution of water throughout the column (Siddique et al. 2010). In geotechnical applications, column tests are generally used to assess the leaching behavior of contaminated soils (Kandsi et al. 2019) and are sometimes performed in parallel with batch tests to compare the results of these different tests for determining contaminant transport parameters in porous media (Naveen et al. 2018; Skjennum et al. 2023; Sumalatha et al. 2022).

Monolithic leaching or mass transfer methods are intended to evaluate the release of elements from a solid specimen and are usually applied to determine the release of contaminants from stabilized materials incorporating waste (Siddique et al. 2010). In these tests, a monolithic sample is immersed in a tank with a leaching fluid without circulation. Method 1315 (USEPA, 2017) allows determining the transfer rates of chemical elements from monolithic or granular materials to different solutions, simulating the leaching behavior of AAM when in contact with natural water, seawater, and moderately acidified water (Zhang et al. 2022). In the adapted method proposed by Bruschi et al.

(2022), the leaching test was performed in parallel with the wetting-drying durability test using distilled water, thus simulating the leaching behavior of the cemented material exposed to weathering.

In summary, the leaching tests presented in Table 1 include the procedures described in Method 1311 (USEPA, 1992); Method 1312 (USEPA, 1994); EN12457 (EN, 2002); HJ/T 299 (MEE, 2007b); HJ/T 300 (MEE, 2007c); D4874 (ASTM, 2014); CEN/TS 14429 (CEN, 2015); EN 14997 (EN, 2015); Method 1315 (USEPA, 2017); and the methods proposed by Bruschi et al. (2022) and Shagñay et al. (2023). The particle size of the cemented material, the solution used, and the test duration varied depending on the chosen method (Table 1). Except for column tests and tests on monolithic materials, in the others, specimens were crushed to a size below 3 cm, 9.5 mm, 4 mm, 3 mm, or 1 mm. Distilled/demineralized water or some acidic solution (i.e., a mixture of water with acetic, sulfuric, or nitric acids) was used as the leaching fluid. Most tests used solid/liquid (S/L) ratios of 1:20 or 1:10. The test durations varied from a few hours to over two months.

Finally, in all these tests, a pattern of using the technique of Inductively Coupled Plasma Optical Emission Spectrometry (ICP-OES) for determining metal concentrations in liquid extracts was observed, in addition to Inductively Coupled Plasma Mass Spectrometry (ICP-MS). The quantified metal concentrations were compared with standards established by different regulations, mainly those that regulate toxicity limits and water quality. The standards and their respective purposes corresponded to: EPA-F-94-001 (USEPA, 1999) - drinking water quality; GB3838 (MEE, 2002) - surface water quality in China; Dutch list (VROM, 2000) - groundwater quality; Council Decision 2003/33/CE (COUNCIL-EU, 2003) - inert and non-hazardous waste; GB5085 (MEE, 2007a) - hazardous waste in China; IWRG 631 (EPA-Victoria, 2009) - hazardous solid waste; CONAMA 460 (CONAMA, 2013) - soil and groundwater quality; Decree 843 (FINLEX, 2017) - reused material in earthworks, and (USEPA, 2022) - drinking water quality. The comparison of leaching responses of elements with standards from regulations is important to understand the advances in metal encapsulation in cemented matrices and to identify or confirm potential applications of stabilized/solidified materials.

Table 1. Leaching tests of alkali-activated materials in the geotechnical literature

Method	Test	Specimen	Solution	S/L	Duration	Reference
Method 1311 (USEPA, 1992)	Solid waste: Leaching toxicity	< 9.5 mm	Acetic acid (pH 2.88) or Acetic acid + sodium hydroxide (pH 4.93)	1:20	18 ± 2 h	Feng et al. (2020); Hoy et al. (2016); Komaei et al. (2023); Xu et al. (2019); Wang et al. (2018)
Method 1312 (USEPA, 1994)	Synthetic precipitation leaching	< 9.5 mm	Sulfuric and nitric acids (pH 4.20)	1:20	18 ± 2 h	Li et al. (2019); Xu et al. (2019)
EN12457 (EN, 2002)	Solid waste: Leaching of granular waste and sludges	< 4 mm	Distilled water	1:10	24 h	Adediran et al. (2021); Cristelo et al. (2021); Ponomar et al. (2023)
NBR 10005 (ABNT, 2004)	Solid waste: Leaching toxicity	< 9.5 mm	Acetic acid + sodium hydroxide (pH 4.93) or Acetic acid (pH 2.88)	1:20	18 ± 2 h	Ferrazzo et al. (2023 a, b); Pelissaro et al. (2023)
HJ/T 299 (MEE, 2007b)	Solid waste: Leaching toxicity	< 9.5 mm	Sulfuric and nitric acids (pH 3.20)	1:10	18 ± 2 h	Pu et al. (2022); Xu et al. (2019)
		< 3 mm	Distilled water	1:10	Oscillated for 8 h and rested for 16 h	Pan et al. (2022)
HJ/T 300 (MEE, 2007c)	Solid waste: Leaching toxicity	< 9.5 mm	Acetic acid (pH 2.64)	1:20	18 ± 2 h	Xu et al. (2019)
CEN/TS 14429 (CEN, 2015)	Solid waste: Leaching behaviour	< 1 mm	Nitric acids (variable pH)	1:10	48 h	Xu et al. (2019)
No specified	Column leaching test	Cylindrical 65 mm x 130 mm	Distilled water	–	10–50 h	Tigue et al. (2018)

Table 1. (Continued)

Method	Test	Specimen	Solution	S/L	Duration	Reference
D4874 (ASTM, 2014)	Leaching in column apparatus	Cylindrical 50 mm x 100 mm	Distilled water	–	hours – weeks	Pereira dos Santos (2021); Ferrazzo et al. (2023b)
D559 (ASTM, 2015) and NEM7375 (NEN, 2004)	Adapted leaching test	Cylindrical 100 mm x 127.3 mm	Distilled water	3L in each tank	3 durability cycles	Bruschi et al. (2022)
Method 1315 (USEPA, 2017)	Static monolithic leaching testing	Cubic	Distilled water, salty water, and acidic water	1 cm^2: 9 mL	0.08–63 days	Zhang et al. (2022)
Garcia-lodeiro et al. (2021)	Accelerated leaching: Decalcification	< 3 cm	6M NH_4NO_3	1 cm^2: 8 cm3	1–28 days	Shagñay et al. (2023)

It is evident that the leaching of AAM has been investigated under various conditions; however, it is limited to the proposed method used. Different tests will provide different estimates of the leaching behavior of AAM. This can also make it difficult to compare results with existing literature. For future research on AAM leaching, it is recommended to use the Leaching Environmental Assessment Framework - LEAF (USEPA, 2014), which provides guidance for defining and using leaching tests for different waste and material types, considering material properties and different environmental conditions in disposal and application scenarios. This document recommends combinations of tests and the substitution of those that are impractical, aiming to obtain a more accurate estimate of leaching behavior among potential uncertainties (e.g., leaching over time and possible changes/influences of environmental conditions).

Final Considerations

The evaluation of environmental risks associated with AAM through metal leaching has become a fundamental factor for the technical viability of materials incorporating waste. In this context, this chapter provides an overview of the leaching behavior of AAM in environmental geotechnics, highlighting key findings from the literature and the specificities of the tests used. The environmental performance of materials stabilized through alkaline activation has been investigated in three main target groups: (i) industrial and agro-industrial solid waste (e.g., asphalt material, ash, slag, and waste foundry sand); (ii) soils contaminated with metals (e.g., Pb, Zn, Cd, and As); and (iii) natural soils (e.g., sandy and clayey soils) and mining waste (e.g., gold ore tailings and bauxite tailings).

Alkaline activation has been responsible for the stabilization/solidification of metals present in the waste used as precursors for alkaline cements and in the materials targeted for treatment, such as reclaimed asphalt pavement, waste foundry sand, metals-contaminated soils, natural soils, and mining tailings. The technique has proven effective in encapsulating or inerting metals in most AAM. In general, AAM showed no metal toxicity, and some even met water quality standards or were suitable for incorporation in earthworks, such as pavements. In some cases, further optimization of alkaline cement in terms of environmental performance, the strategies/dosages of alkaline cement applied to the stabilization of waste, soils, or mining residues, and the understanding of the mechanisms of inerting problematic metals may still be necessary.

The leaching behavior of AAM was evaluated through batch tests, column tests, and monolithic sample tests with different procedures and purposes. The literature review revealed that the leaching of AAM was primarily assessed through batch tests developed for waste classification/characterization. Different tests provide different estimates of the leaching behavior of AAM, meaning that materials with specific properties are being evaluated under various scenarios. This can make it difficult to compare results from the literature and underscores the need for international standardization of leaching tests for materials that combine waste and alkaline activation. The proposal of the USEPA's LEAF (2014) is suggested as a tool for investigating the environmental performance of AAM in future research.

This chapter contributes to a comprehensive understanding of AAM leaching in environmental geotechnics, including the types of materials studied, behavior and identified mechanisms, as well as the specificities of the tests and methods used. It was possible to identify potentialities, challenges, unexplored research gaps in the literature, and avenues for the evaluation of AAM leaching. It is expected that this information will serve as a reference for future studies on AAM, aiming to overcome the primary barrier to applications, namely, ensuring that the material poses no environmental risks.

References

ABNT – Associação Brasileira de Normas Técnicas. (2004). *Procedimento para obtenção de extrato lixiviado de resíduos sólidos*. (ABNT NBR10005: 20004). ABNT. https://www.normas.com.br/.

Adediran, A., Yliniemi, J. & Illikainen, M. (2021). Mineralogy and glass content of Fe-rich fayalite slag size fractions and their effect on alkali activation and leaching of heavy metals. *Ceramic Engineering & Science*, 1–14. https://doi.org/10.1002/ces2.10107.

Ahmari, S. & Zhang, L. (2013). Durability and leaching behavior of mine tailings-based geopolymer bricks. *Construction and Building Materials*, *44*, 743–750. https://doi.org/10.1016/j.conbuildmat.2013.03.075.

ASTM – American Society for Testing and Materials. (2014). *Standard test method for leaching solid material in a column apparatus*. (ASTM D4874: 2014). ASTM International. https://www.astm.org/.

ASTM – American Society for Testing and Materials. (2015). *Standard test methods for wetting and drying compacted soil-cement mixtures*. (ASTM D559/D559M: 2015). ASTM International. https://www.astm.org/.

ASTM – American Society for Testing and Materials. (2016). *Standard test methods for measurement of hydraulic conductivity of saturated porous materials using a flexible wall oermeameter.* (ASTM D5084: 2016). ASTM International. https://www.astm.org/.

Bruschi, G. J., Dos Santos, C. P., Levandoski, W. M. K., Ferrazzo, S. T., Korf, E. P., Saldanha, R. B. & Consoli, N. C. (2022). Leaching assessment of cemented bauxite tailings through wetting and drying cycles of durability test. *Environmental Science and Pollution Research*, *29*, 59247–59262. https://doi.org/10.1007/s11356-022-20031-5.

CONAMA – Conselho Nacional Do Meio Ambiente. (2013). *Altera a Resolução no 420, de 28 de dezembro de 2009, do Conselho Nacional do Meio Ambiente-CONAMA, que dispõe sobre critérios e valores orientadores de qualidade do solo quanto à presença de substâncias químicas e dá outras providências*. (Resolução CONAMA n° 460: 2013). CONAMA. http://conama.mma.gov.br/.

COUNCIL-EU. (2003). Council Decision 2003/33/EC of 19 December 2002. *Official Journal of the European Communities*, *16*(11), 27–49.

Cristelo, N., Rivera, J., Miranda, T. & Fernández-Jiménez, A. (2021). Stabilisation of a plastic soil with alkali activated cements developed from industrial wastes. *Sustainability*, *13*, 4501. https://doi.org/https://doi.org/10.3390/su13084501.

Desmarais, T. L. & Costa, M. (2019). Mechanisms of chromium-induced toxicity. *Current Opinion in Toxicology*, *14*, 1–7. https://doi.org/10.1016/j.cotox.2019.05.003.

EN – European Standards. (2002). *Characterisation of waste. Leaching. Compliance test for leaching of granular waste materials and sludges. One stage batch test at a liquid to solid ratio of 10 l/kg for mate- rials with particle size below 4 mm (without or with size reduction).* (BS EN12457-2: 2002). EN. https://www.en-standard.eu/.

CEN – European Committee for Standardization. (2015). *Characterization of waste - Leaching behaviour test - Influence of pH on leaching with initial acid/base addition*. (CEN/TS 14429: 2015). CEN. https://www.cencenelec.eu/.

EN – European Standards. (2015). *Characterization of waste. Leaching behaviour test. Influence of pH on leaching with continuous pH control*. (EN 14997: 2015). EN. https://www.en-standard.eu/.

EPA-Victoria – Environmental Protection Agency of Victoria. (2009). *Solid industrial waste hazard categorization and management, industrial waste resource guidelines*. (IWRG 631: 2009). EPA-Victoria. https://www.epa.vic.gov.au/.

Feng, Y. S., Zhou, S. J., Xia, W. Y. & Du, Y. J. (2020). Solidify/stabilise a heavy metal-contaminated soil using a novel steel slag-based binder. *Environmental Geotechnics*, 1–16. https://doi.org/https://doi.org/10.1680/jenge.19.00226.

Ferrazzo, S. T., Tonini de Araújo, M., Bruschi, G. J., Chaves, H. M., Korf, E. P. & Consoli, N. C. (2023a). Mechanical and environmental behavior of waste foundry sand stabilized with alkali-activated sugar cane bagasse ash-eggshell lime binder. *Construction and Building Materials*, *383*, 131313. https://doi.org/https://doi.org/10.1016/j.conbuildmat.2023.131313.

Ferrazzo, S. T., Tonini de Araújo, M., Bruschi, G. J., Korf, E. P., Levandoski, W. M. K., Pereira dos Santos, C. & Consoli, N. C. (2023b). Metal encapsulation of waste foundry sand stabilized with alkali-activated binder: Batch and column leaching tests. *Journal of Environmental Management, 348*, 119287. https://doi.org/10.1016/j.jenvman.2023.119287.

FINLEX. (2017). Government Decree on the Recovery of Certain Wastes in Earth Construction (843/2017). *FINLEX*, 1–14.

Garcia-lodeiro, I., Goracci, G., Dolado, J. S. & Blanco-varela, M. T. (2021). Mineralogical and microstructural alterations in a portland cement paste after an accelerated decalcification process. *Cement and Concrete Research, 140*, 106312. https://doi.org/10.1016/j.cemconres.2020.106312.

Ha, F., Zhang, H., Li, Y. & Zhang, Z. (2023). Recycling and comprehensive utilization of ferronickel slag in concrete. *Journal of Cleaner Production, 414*, 137633. https://doi.org/https://doi.org/10.1016/j.jclepro.2023.137633.

Han, C. & Hong, Y. (2018). Adverse health effects of ferronickel manufacturing factory on local residents: An interrupted time series analysis. *Environment International, 114*, 288–296. https://doi.org/10.1016/j.envint.2018.02.045.

Hoy, M., Horpibulsuk, S., Rachan, R., Chinkulkijniwat, A. & Arulrajah, A. (2016). Recycled asphalt pavement – fly ash geopolymers as a sustainable pavement base material: Strength and toxic leaching investigations. *Science of the Total Environment, 573*, 19–26. https://doi.org/10.1016/j.scitotenv.2016.08.078.

Jiang, Q., He, Y., Wu, Y., Dian, B., Zhang, J., Li, T. & Jiang, M. (2022). Solidification/stabilization of soil heavy metals by alkaline industrial wastes: A critical review. *Environmental Pollution, 312*, 120094. https://doi.org/https://doi.org/10.1016/j.envpol.2022.120094.

Kandsi, B., Benhabib, K., Mimanne, G., Djellouli, M. & Taleb, S. (2019). Assessment of the potential mobility of copper in contaminated soil samples by column leaching test. *Eurasian Journal of Soil Science, 8*, 27–34. https://doi.org/10.18393/ejss.485939.

Kiventerä, J., Sreenivasan, H., Cheeseman, C., Kinnunen, P. & Illikainen, M. (2018). Immobilization of sulfates and heavy metals in gold mine tailings by sodium silicate and hydrated lime. *Journal of Environmental Chemical Engineering, 6*(5), 6530–6536. https://doi.org/10.1016/j.jece.2018.10.012.

Komaei, A., Noorzad, A. & Ghadir, P. (2023). Stabilization and solidification of arsenic contaminated silty sand using alkaline activated slag. *Journal of Environmental Management, 344*, 118395. https://doi.org/https://doi.org/10.1016/j.jenvman.2023.118395.

Kumar, A., Kumar, A., M.M.S., C. P., Chaturvedi, A. K., Shabnam, A. A., Subrahmanyam, G., Mondal, R., Gupta, D. K., Malyan, S. K., Kumar, S. S., Khan, S. A. & Yadav, K. K. (2020). Lead toxicity: health hazards, influence on food chain, and sustainable remediation approaches. *International Journal of Environmental Research and Public Health, 17*, 2179. https://doi.org/10.3390/ijerph17072179.

Li, Y., Zhang, T., Jia, S., Liu, J., Quan, X. & Zheng, W. (2019). Mechanical properties and leaching characteristics of geopolymer-solidified/stabilized lead-contaminated soil. *Advances in Civil Engineering, 2019*, 1–8. https://doi.org/https://doi.org/10.1155/2019/6015769.

Liu, J., Zha, F., Deng, Y., Cui, K. & Zhang, X. (2017). Effect of an alkaline environment on the engineering behavior of cement-stabilized/solidified Zn-contaminated soils.

Environmental Science and Pollution Research, *24*, 28248–28257. https://doi.org/10.1007/s11356-017-0400-9.

Luo, X., Wu, C., Lin, Y., Li, W., Deng, M., Tan, J. & Xue, S. (2023). Soil heavy metal pollution from Pb/Zn smelting regions in China and the remediation potential of biomineralization. *Journal of Environmental Sciences*, *125*, 662–677. https://doi.org/https://doi.org/10.1016/j.jes.2022.01.029.

MEE – Ministry of Ecology and Environment the People's Republic of China. (2002). *Environmental quality standard for surface water*. (MEE GB3838: 2002). MEE. https://english.mee.gov.cn/.

MEE – Ministry of Ecology and Environment the People's Republic of China. (2007a). *Identification standards for hazardous wastes-identification for extraction toxicity*. (MEE GB5085: 2007). MEE. https://english.mee.gov.cn/.

MEE – Ministry of Ecology and Environment the People's Republic of China. (2007b). *Solid waste-extraction procedure for leaching toxicity–sulphuric acid & nitric acid method*. (MEE HJ/T299: 2007). MEE. https://english.mee.gov.cn/.

MEE – Ministry of Ecology and Environment the People's Republic of China. (2007c). *Solid waste-extraction procedure for leaching toxicity–acetic acid buffer solution method*. (MEE HJ/T300: 2007). MEE. https://english.mee.gov.cn/.

Movaheditabar, P., Javaherian, M. & Nobakht, V. (2022). Steelmaking slag as a valuable industrial waste in the preparation of 5-substituted-1H-tetrazoles on water. *Journal of Sustainable Metallurgy*, *8*, 882–892. https://doi.org/10.1007/s40831-022-00533-9.

Naka, A., Yasutaka, T., Sakanakura, H., Kalbe, U., Watanabe, Y., Inoba, S., Takeo, M., Inui, T., Katsumi, T., Fujikawa, T., Sato, K., Higashino, K. & Someya, M. (2016). Column percolation test for contaminated soils: Key factors for standardization. *Journal of Hazardous Materials*, *320*, 326–340. https://doi.org/10.1016/j.jhazmat.2016.08.046.

Naveen, B. P., Sumalatha, J. & Malik, R. K. (2018). A study on contamination of ground and surface water bodies by leachate leakage from a landfill in Bangalore, India. *International Journal of Geo-Engineering*, *9*(27). https://doi.org/10.1186/s40703-018-0095-x.

NEN – Netherlands Standards. (2004). *Leaching characteristics - Determination of the leaching of inorganic components from moulded or monolitic materials with a diffusion test - solid earthy and stony materials*. (NEN 7375: 2004). NEN. https://www.nen.nl/en.

Ou, X., Chen, S., Jiang, J., Qin, J. & Zhang, L. (2022). Reuse of red mud and bauxite tailings mud as subgrade materials from the perspective of mechanical properties. *Materials*, *15*, 1123. https://doi.org/https://doi.org/10.3390/ma15031123.

Pan, Z., Zhang, C., Li, Y. & Yang, C. (2022). Solidification/stabilization of gold ore tailings powder using sustainable waste-based composite geopolymer. *Engineering Geology*, *309*, 106793. https://doi.org/10.1016/j.enggeo.2022.106793.

Pelissaro, D. T., Dalla Rosa, F. & Korf, E. P. (2023). Durability, long-term and environmental evaluation of alkali-activated alternative soluble silica source for recycled asphalt pavement stabilization. *Journal of Materials in Civil Engineering*. https://doi.org/10.1061/JMCEE7/MTENG-16168.

Pereira dos Santos, C. (2021). *Estabilização de rejeito de mineração de ouro por meio da álcali-ativação de cal de carbureto e cinza do bagaço de cana-de-açúcar*. Universidade Federal do Rio Grande do Sul.

Ponomar, V., Ohenoja, K. & Illikainen, M. (2023). Optimizing activating solution and environmental leaching characteristics of Fe-rich alkali-activated Zn slag. *Journal of Hazardous Materials*, *445*, 130575. https://doi.org/10.1016/j.jhazmat.2022.130575.

Pu, S., Duan, W., Zhu, Z., Wang, W., Zhang, C., Li, N., Jiang, P. & Wu, Z. (2022). Environmental behavior and engineering performance of self-developed soil silico-aluminophosphate geopolymer binder stabilized lead contaminated soil. *Journal of Cleaner Production*, *379*, 134808. https://doi.org/https://doi.org/10.1016/j.jclepro.2022.134808.

Rahman, S., Reza, A. H. M. S., Siddique, A. B., Akbor, A. & Hasan, M. (2023). Accumulation of arsenic and other metals in soil and human consumable foods of Meherpur district, southwestern Bangladesh, and associated health risk assessment. *Environmental Sciences Europe*, *35*(47). https://doi.org/10.1186/s12302-023-00751-2.

Saldanha, R. B., Caicedo, A. M. L., De Araújo, M. T., Scheuermann Filho, H. C., Moncaleano, C. J., Silvia, J. P. S. & Consoli, C. (2023). Potential use of iron ore tailings for binder production: A life cycle assessment. *Construction and Building Materials*, *365*, 130008. https://doi.org/https://doi.org/10.1016/j.conbuildmat.2022.130008.

Servi, S., Lotero, A., Silva, J. P. S., Bastos, C. & Consoli, N. C. (2022). Mechanical response of filtered and compacted iron ore tailings with different cementing agents: Focus on tailings-binder mixtures disposal by stacking. *Construction and Building Materials*, *349*, 128770. https://doi.org/https://doi.org/10.1016/j.conbuildmat.2022.128770.

Shagñay, S., Garcia-lodeiro, I., Velasco, F., Bautista, A. & Torres-carrasco, M. (2023). Mineralogical and microstructural changes in alkali-activated and hybrid materials exposed to accelerated leaching. *Journal of Building Engineering*, *64*, 105733. https://doi.org/10.1016/j.jobe.2022.105733.

Siddique, R., Kaur, G. & Rajor, A. (2010). Waste foundry sand and its leachate characteristics. *Resources, Conservation & Recycling*, *54*(12), 1027–1036. https://doi.org/10.1016/j.resconrec.2010.04.006.

Skjennum, K. A., French, H. K., Carotenuto, P. & Okkenhaug, G. (2023). Combined column test for characterization of leaching and transport of trace elements in contaminated soils. *Water*, *15*, 874. https://doi.org/10.3390/w15050874.

Sumalatha, J., Kumar, C. L. M., Prashant, S. & Shwetha, K. G. (2022). Design of soil remediation techniques from column leaching test results. *International Journal of Engineering, Transactions B: Applications*, *35*(2), 344–350. https://doi.org/10.5829/ije.2022.35.02b.10.

Tang, L., Tang, C., Xiao, J., Zeng, P. & Tang, M. (2018). A cleaner process for valuable metals recovery from hydrometallurgical zinc residue. *Journal of Cleaner Production*, *201*, 764–773. https://doi.org/10.1016/j.jclepro.2018.08.096.

Tigue, A. A. S., Malenab, R. A. J., Dungca, J. R., Yu, D. E. C. & Promentilla, M. A. B. (2018). Chemical stability and leaching behavior of one-part geopolymer from soil and coal fly ash mixtures. *Minerals*, *8*(411). https://doi.org/10.3390/min8090411.

USEPA – United States Environmental Protection Agency. (1992). Method 1311: Toxicity Characteristic Leaching Procedure. In *SW-846: Hazardous Waste Test Methods* (1st ed., pp. 1–35). USEPA. https://www.epa.gov/sites/default/files/2015-12/documents/1311.pdf.

USEPA – United States Environmental Protection Agency. (1994). Method 1312: Synthetic Precipitation Leaching Procedure. In *SW-846: Hazardous Waste Test Methods* (1st ed., pp. 1–30). USEPA.

USEPA – United States Environmental Protection Agency. (1999). EPA-F-94-001: National Primary Drinking Water Standards. *USEPA*.

USEPA – United States Environmental Protection Agency. (2014). *Leaching test relationships, laboratory-to-field comparisons and recommendations for leaching evaluation using the Leaching Environmental Assessment Framework (LEAF)* (D. S. Kosson, H. A. van der Sloot, A. C. Garrabrants, & P. F. A. B. Seignette (eds.); 1st ed.). USEPA.

USEPA – United States Environmental Protection Agency. (2017). Method 1315: Mass transfer rates of constituents in monolithic or compacted granular materials using a semi-dynamic tank leaching procedure. In *SW-846: Hazardous Waste Test Methods* (Vol. 7345, pp. 1–37). USEPA.

USEPA – United States Environmental Protection Agency. (2022). *Ground Water and Drinking Water: National Primary Drinking Water Regulations*. USEPA. https://www.epa.gov/ground-water-and-drinking-water/national-primary-drinking-water-regulations#Inorganic.

VROM – Ministerie van Volksuisveting Ruimtelijke Ordening En Milieubeheer. (2000). Dutch Target and Intervention Values. *Ministerie van Volksuisveting Ruimtelijke Ordening En Milieubeheer*, 1–51.

Wang, D., Wang, Q., Zhuang, S. & Yang, J. (2018). Evaluation of alkali-activated blast furnace ferronickel slag as a cementitious material: Reaction mechanism, engineering properties and leaching behaviors. *Construction and Building Materials*, *188*, 860–873. https://doi.org/10.1016/j.conbuildmat.2018.08.182.

Wani, A. L., Ara, A. & Usmani, J. A. (2015). Lead toxicity: a review. *Interdiciplinary Toxicology*, *8*(2), 55–64. https://doi.org/10.1515/intox-2015-0009.

Wei, M., Du, Y., Reddy, K. R. & Wu, H. (2015). Effects of freeze-thaw on characteristics of new KMP binder stabilized Zn- and Pb-contaminated soils. *Environmental Science and Pollution Research*, *22*, 19473–19484. https://doi.org/10.1007/s11356-015-5133-z.

Xu, P., Zhao, Q., Qiu, W. & Xue, Y. (2019). The evaluation of the heavy metal leaching behavior of MSWI-FA added alkali-activated materials bricks by using different leaching test methods. *International Journal of Environmental Research and Public Health*, *16*, 1151. https://doi.org/10.3390/ijerph16071151.

Zhang, N., Hedayat, A., Figueroa, L., Steirer, K. X., Li, H., Sosa, H. G. B., Bernal, R. P. H., Tupa, N., Morales, I. Y. & Loza, R. S. C. (2022). Experimental studies on the durability and leaching properties of alkali-activated tailings subjected to different environmental conditions. *Cement and Concrete Composites*, *130*, 104531. https://doi.org/10.1016/j.cemconcomp.2022.104531.

About the Editors

Eduardo Pavan Korf
Laboratory of Solid Wastes and Environmental Geotechnics
Federal University of Fronteira Sul
Erechim, RS, Brazil
E-mail: eduardo.korf@uffs.edu.br

Pedro Domingos Marques Prietto
Graduate Program in Civil and Environmental Engineering
University of Passo Fundo
Passo Fundo, RS, Brazil
E-mail: pdmp@upf.br

Caroline Müller
Graduate Program in Environmental Science and Technology
Federal University of Fronteira Sul
Erechim, RS, Brazil
E-mail: carolinemulleram@gmail.com

Suéllen Tonatto Ferrazzo
Laboratory of Geotechnical and Geoenvironmental Engineering
Federal University of Rio Grande do Sul
Porto Alegre, RS, Brazil
E-mail: suellenferrazzo@hotmail.com

Index

R

S

T

V

W

X

Z